AF361886

Marine Engines Performance and Emissions II

Marine Engines Performance and Emissions II

Editor

María Isabel Lamas Galdo

MDPI • Basel • Beijing • Wuhan • Barcelona • Belgrade • Manchester • Tokyo • Cluj • Tianjin

Editor
María Isabel Lamas Galdo
Universidade da Coruña
Ferrol
Spain

Editorial Office
MDPI
St. Alban-Anlage 66
4052 Basel, Switzerland

This is a reprint of articles from the Special Issue published online in the open access journal *Journal of Marine Science and Engineering* (ISSN 2077-1312) (available at: https://www.mdpi.com/journal/jmse/special_issues/marine_engines_performance_emissions_II).

For citation purposes, cite each article independently as indicated on the article page online and as indicated below:

LastName, A.A.; LastName, B.B.; LastName, C.C. Article Title. *Journal Name* **Year**, *Volume Number*, Page Range.

ISBN 978-3-0365-7992-4 (Hbk)
ISBN 978-3-0365-7993-1 (PDF)

Contents

About the Editor

María Isabel Lamas Galdo

María Isabel Lamas Galdo has a Ph.D. in Industrial Engineering. She has worked as an associate professor at the Higher Polytechnic University College of Universityof Coruña, Spain, since 2008. Moreover, she has collaborated with the Norplan Engineering S.L. Her main interest is in computational fluid dynamics (CFD), specially applied to thermal systems, marine engines, and pollution reduction. She has written several books. She is the author of dozens of publications in ISI journals and she has participated in several national and international conferences and research projects. She also has experience in engineering projects.

Preface to "Marine Engines Performance and Emissions II"

Marine engines are efficient machines, but stricter legislation about emissions makes continuous research necessary in order to reduce their consumption and improve their efficiency. This Special Issue, "Marine Engines Performance and Emissions II", is a continuation of "Marine Engines Performance and Emissions". It contains a set of peer-reviewed works about marine engine performance and emissions.

María Isabel Lamas Galdo
Editor

Journal of
*Marine Science
and Engineering*

Editorial

Marine Engines Performance and Emissions II

María Isabel Lamas Galdo

Escuela Politécnica de Ingeniería de Ferrol, CITENI, Campus Industrial de Ferrol, Universidade da Coruña, Mendizábal, 15403 Ferrol, Spain; isabel.lamas.galdo@udc.es

Citation: Lamas Galdo, M.I. Marine Engines Performance and Emissions II. *J. Mar. Sci. Eng.* **2022**, *10*, 1987. https://doi.org/10.3390/jmse10121987

Received: 30 November 2022
Accepted: 8 December 2022
Published: 14 December 2022

Publisher's Note: MDPI stays neutral with regard to jurisdictional claims in published maps and institutional affiliations.

Engines are one of the most important components of ships. The performance of marine engines has been consecutively improved over the years, and current marine engines are more efficient year by year. Regarding emissions, marine engines have also been modified over the years. Nevertheless, marine pollution constitutes an important contribution worldwide. The scientific community is investing significant effort in developing efficient and less pollutant marine engines. According to this, the Special Issue, "Marine Engines Performance and Engines", published as a book [1], was prepared to collect works relating to marine engine performance and emissions in general. A total of 12 works [2–13] were published in this first Special Issue. In order to continue this work, the present Special Issue, "Marine Engines Performance and Engines II", was developed to include more works related to this topic, such as emissions from marine engines, after-treatments, conventional and alternative fuels, mathematical models, marine engine technology, combustion, design, control, injection, lubrication and lubes, auxiliary systems, transport, etc. A total of nine works were published in this second Special Issue.

A special interest was focused on new technologies. Particularly, Markovič et al. [14] analyzed a new generation of the compact system for performing measurements of sold liquids by gas station dispensers; Jírová et al. [15] analyzed an original vibrodiagnostic device to control linear rolling conveyor reliability; Živčák et al. [16] analyzed how to increase the mechanical properties of 3D printed samples by direct metal laser sintering using heat treatment processes; and Varbanets et al. [17], who studied an acoustic method for the estimation of marine low-speed engine turbocharger parameters. Another important topic broached in this Special Issue was related to emissions from marine engines. In this regard, Kao et al. [18] analyzed an AIS-based scenario simulation for the control and improvement of ship emissions in ports. Regarding emission reduction by using alternative fuels, Rodríguez et al. [19] analyzed the possibilities of ammonia as both fuel and NOx reductant in marine engines, and Jablonický et al. [20] developed an assessment of the technical and ecological parameters of a diesel engine in the application of new samples of biofuels. Regarding emissions reduction using post-treatment processes, Ryu and Park [21] analyzed a composite scrubber with a built-in silencer for marine engines. Finally, numerical methods were treated on the works of Rodriguez et al. [19] and Maláková et al. [22].

Conflicts of Interest: The author declares no conflict of interest.

References

1. Lamas Galdo, M.I. Marine Engines Performance and Emissions. *J. Mar. Sci. Eng.* **2021**, *9*, 280. [CrossRef]
2. Puškár, M.; Kopas, M.; Sabadka, D.; Kliment, M.; Šoltésová, M. Reduction of the Gaseous Emissions in the Marine Diesel Engine Using Biodiesel Mixtures. *J. Mar. Sci. Eng.* **2020**, *8*, 330. [CrossRef]
3. Sui, C.; de Vos, P.; Stapersma, D.; Visser, K.; Ding, Y. Fuel Consumption and Emissions of Ocean-Going Cargo Ship with Hybrid Propulsion and Different Fuels over Voyage. *J. Mar. Sci. Eng.* **2020**, *8*, 588. [CrossRef]
4. Perez, J.R.; Reusser, C.A. Optimization of the Emissions Profile of a Marine Propulsion System Using a Shaft Generator with Optimum Tracking-Based Control Scheme. *J. Mar. Sci. Eng.* **2020**, *8*, 221. [CrossRef]
5. Winnes, H.; Fridell, E.; Moldanová, J. Effects of Marine Exhaust Gas Scrubbers on Gas and Particle Emissions. *J. Mar. Sci. Eng.* **2020**, *8*, 299. [CrossRef]
6. Kim, K.-H.; Kong, K.-J. One-Dimensional Gas Flow Analysis of the Intake and Exhaust System of a Single Cylinder Diesel Engine. *J. Mar. Sci. Eng.* **2020**, *8*, 1036. [CrossRef]
7. Lamas Galdo, M.I.; Castro-Santos, L.; Rodriguez Vidal, C.G. Numerical Analysis of NOx Reduction Using Ammonia Injection and Comparison with Water Injection. *J. Mar. Sci. Eng.* **2020**, *8*, 109. [CrossRef]
8. Witkowski, K. Research of the Effectiveness of Selected Methods of Reducing Toxic Exhaust Emissions of Marine Diesel Engines. *J. Mar. Sci. Eng.* **2020**, *8*, 452. [CrossRef]
9. Lehtoranta, K.; Koponen, P.; Vesala, H.; Kallinen, K.; Maunula, T. Performance and Regeneration of Methane Oxidation Catalyst for LNG Ships. *J. Mar. Sci. Eng.* **2021**, *9*, 111. [CrossRef]
10. Lamas, M.I.; Castro-Santos, L.; Rodriguez, C.G. Optimization of a Multiple Injection System in a Marine Diesel Engine through a Multiple-Criteria Decision-Making Approach. *J. Mar. Sci. Eng.* **2020**, *8*, 946. [CrossRef]
11. Homišin, J.; Kaššay, P.; Urbanský, M.; Puškár, M.; Grega, R.; Krajňák, J. Electronic Constant Twist Angle Control System Suitable for Torsional Vibration Tuning of Propulsion Systems. *J. Mar. Sci. Eng.* **2020**, *8*, 721. [CrossRef]
12. Shen, H.; Zhang, J.; Yang, B.; Jia, B. Development of a Marine Two-Stroke Diesel Engine MVEM with In-Cylinder Pressure Trace Predictive Capability and a Novel Compressor Model. *J. Mar. Sci. Eng.* **2020**, *8*, 3020. [CrossRef]
13. Píštěk, V.; Kučera, P.; Fomin, O.; Lovska, A. Effective Mistuning Identification Method of Integrated Bladed Discs of Marine Engine Turbochargers. *J. Mar. Sci. Eng.* **2020**, *8*, 379. [CrossRef]
14. Markovič, J.; Živčák, J.; Sága, M.; Tarbajovský, P. New generation of the compact system for performing measurements of solid liquids by gas station dispensers. *J. Mar. Sci. Eng.* **2022**, *10*, 524. [CrossRef]
15. Jírová, R.; Pešík, L.; Grega, R. An Original Vibrodiagnostic Device to Control Linear Rolling Conveyor Reliability. *J. Mar. Sci. Eng.* **2022**, *10*, 445. [CrossRef]
16. Živčák, J.; Nováková-Marcinčínová, E.; Nováková-Marcinčínová, L.; Balint, T.; Puškár, M. Increasing Mechanical Properties of 3D Printed Samples by Direct Metal Laser Sintering Using Heat Treatment Process. *J. Mar. Sci. Eng.* **2021**, *9*, 821. [CrossRef]
17. Varbanets, R.; Fomin, O.; Píštěk, V.; Klymenko, V.; Minchev, D.; Khrulev, A.; Zalozh, V.; Kučera, P. Acoustic Method for Estimation of Marine Low-Speed Engine Turbocharger Parameters. *J. Mar. Sci. Eng.* **2021**, *9*, 321. [CrossRef]
18. Kao, S.-L.; Chung, W.-H.; Chen, C.-W. AIS-Based Scenario Simulation for the Control and Improvement of Ship Emissions in Ports. *J. Mar. Sci. Eng.* **2022**, *10*, 129. [CrossRef]
19. Rodríguez, C.G.; Lamas, M.I.; Rodríguez, J.d.D.; Abbas, A. Possibilities of Ammonia as Both Fuel and NOx Reductant in Marine Engines: A Numerical Study. *J. Mar. Sci. Eng.* **2022**, *10*, 43. [CrossRef]
20. Jablonický, J.; Feriancová, P.; Tulík, J.; Hujo, L.; Tkáč, Z.; Kuchar, P.; Tomić, M.; Kaszkowiak, J. Assessment of Technical and Ecological Parameters of a Diesel Engine in the Application of New Samples of Biofuels. *J. Mar. Sci. Eng.* **2022**, *10*, 1. [CrossRef]
21. Ryu, M.-R.; Park, K. Analysis of Composite Scrubber with Built-In Silencer for Marine Engines. *J. Mar. Sci. Eng.* **2021**, *9*, 962. [CrossRef]
22. Maláková, S.; Puškár, M.; Frankovský, P.; Sivák, S.; Harachová, D. Influence of the Shape of Gear Wheel Bodies in Marine Engines on the Gearing Deformation and Meshing Stiffness. *J. Mar. Sci. Eng.* **2021**, *9*, 1060. [CrossRef]

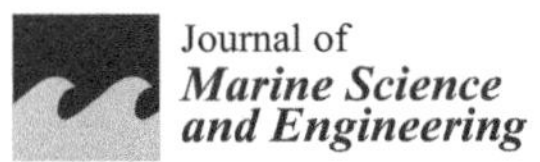

Journal of
*Marine Science
and Engineering*

MDPI

Article

New Generation of the Compact System for Performing Measurements of Sold Liquids by Gas Station Dispensers

Jaromír Markovič [1,*], Jozef Živčák [2], Milan Sága [1] and Pavol Tarbajovský [2]

[1] Faculty of Mechanical Engineering, University of Žilina, Univerzitná 8215/1, 010 26 Zilina, Slovakia; milan.saga@fstroj.uniza.sk

[2] Faculty of Mechanical Engineering, TU Košice, Letná 9, 040 01 Košice, Slovakia; jozef.zivcak@tuke.sk (J.Ž.); pavol.tarbajovsky@tuke.sk (P.T.)

* Correspondence: jaromir.markovic@fstroj.uniza.sk

Abstract: This paper presents an advanced compact system that represents an innovative solution determined for the precise measurement of the fuel liquids and chemical materials used in the transport area. This system was created as a product of the applied research, development, and following realisation in the Slovak Legal Metrology. This organisation is authorised for the certification of the measuring systems. The given system enables to perform metrological control for a wide range of the measuring systems in order to achieve reliable results in the measuring process and the required measuring precision, as well as to minimise idle times of the measuring stands during the metrological control. In addition, the presented system is user-friendly and it requires only a short time for training of the metrological personnel.

Keywords: compact system; sold liquids; metrology; gas station dispensers

Citation: Markovič, J.; Živčák, J.; Sága, M.; Tarbajovský, P. New Generation of the Compact System for Performing Measurements of Sold Liquids by Gas Station Dispensers. *J. Mar. Sci. Eng.* **2022**, *10*, 524. https://doi.org/10.3390/jmse10040524

Academic Editor: María Isabel Lamas Galdo

Received: 21 March 2022
Accepted: 7 April 2022
Published: 10 April 2022

Publisher's Note: MDPI stays neutral with regard to jurisdictional claims in published maps and institutional affiliations.

1. Introduction

Dispensers used for volume measurements of sold liquids at pump stations can be sorted as measuring systems for the continuous and dynamic measurement of amount of liquid, except water. Measurement systems' metrological and technical requirements for continuous and dynamic liquid amount measurements except water, which undergo legal metrological control, are stated in international recommendations OIML R 117-1 [1,2]. International recommendations contained in OIML R 117-1 are used as a normative document within conformity assessment of dispensers offered and in the EU market according to Regulation 2014/32/EU for measuring instruments [3]. OIML R 117-1 includes metrological and technical requirements for diesel, gasoline, LPG, AdBlue, washer fluid dispensers, etc.

Automation is becoming commonly used in industry practice. Automation benefits extend to the field of metrology in the form of rejuvenation for often repetitive processes and are supported by the credibility of measurement results. Automation trends penetrate all industrial branches including metrology. The idea of automation application in the field of metrology control of gauges used at pump stations comes from practice requirements and activities performed by the Research and Development department of Slovak Legal Metrology. Slovak Legal Metrology as the main partner of this project has cooperated with prominent Slovak technical universities.

The main purpose of this article is to introduce a new generation of the compact system for performing metrological controls (mainly confirmation) of dispensers used at pump stations (AMSV). Dispensers are sorted as dynamic measurement systems for liquids except for water. There are several ancestors of this system which differ mainly by the considerably higher level of automation and wider range of provided services [4].

The goal of the project was to create a measurement system that would be able to check all dispensers available at pump stations with the usage of artificial intelligence elements, such as the automation and acceleration of measurement processes with compact construction with the ability of integration into conventional commercial vehicles up to 3.5 tones.

During the design process, the usage of electric systems in a potentially explosive environment was taken into account concerning the safe use of the AMSV system.

Figure 1 shows the AMSV system integrated into a commercial vehicle, thereby creating a mobile system for measurement. Individual subsystems are described in the following sections.

Figure 1. The AMSV system as a part of the conventional commercial vehicle up to 3.5 tones.

2. Mechanical Part of the AMSV System

It was necessary to consider the whole system weight, taking into account its strength and rigidity during the design process. The development of the compact system, which weighs with an appropriate vehicle and whose driver would not exceed the maximum capacity load (3.5 t) of the commercial vehicle, was the goal of this project. In addition, easy insertion of the whole system into the commercial vehicle cargo space was required without modifying vehicles in a wide range. Such a complex system requires a lot of sensors, tanks, standard containers, pumps, hose systems, etc., for the verification of all types of measuring sets. Due to mentioned reasons, it was not possible to use conventional materials such as steel for all construction elements. The best weight-saving parts, as it turned out, were collection tanks and a system of supporting frames. There are three types of frames: mainframe (Figure 2), frame for LPG technology, and frame for standard measuring containers. Collection tanks and the supportive frame system were made out of aluminum alloy, which has one-third density and one-half the failure strength counter to conventionally used steels. The weight of the whole construction was lowered by 350 kg thanks to this approach. Strength analysis of supporting frames and collection tanks was performed in the form of engineering simulations due to safety reasons. The aim of these simulations was:

- The static load of the system under the action of gravitational force, important during the opening process of full collection tanks.
- The biaxial static load of the system under the action of gravitational force and braking force (proportional to gravitational acceleration 0.8 g).
- The biaxial static load of the system under the action gravitational force and centrifugal force (proportional to gravitational acceleration 0.4 g), considering system behavior at curves.

Simulation results of all measurements proved high coefficients of safety (values higher than 4) [5,6].

A 3D model of the system as a compact device is shown in Figure 3.

Figure 2. The AMSV compact system (3D model).

Figure 3. The AMSV compact system (3D model).

3. Electrical Solution and Software Solution

The electrical system is designed to be independent. The power supply system uses four batteries with a total capacity of 200 Ah at their 24 V power supply. The power supply system includes converters, batteries, and diode switches [7–9].

When powered by an external power network, a 230 V diode switch switches the power supply from the batteries to the invertors and the batteries start to charge. The whole electrical system was designed to cover the 15 h activity of the AMSV system at an overdraft of 6000 L of liquid (cabin temperature of 20 °C) or 15 h of activity at overdraft of 4500 L of liquid (cabin temperature of 0 °C). A simplified system electrical scheme is shown in Figure 4.

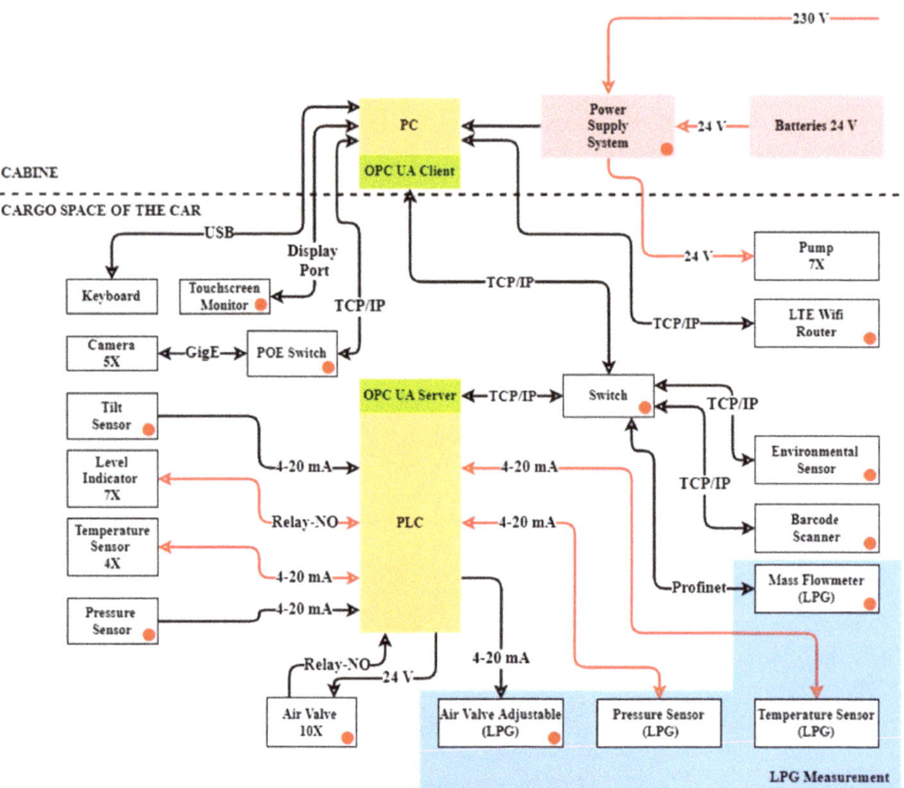

Figure 4. Simplified scheme of electrical system.

The AMSV system is controlled by the technical operator with the help of software, which is run on the computer and together with PLC [10–12] forms the management part at the highest level. The OPC UA communication protocol [13,14] serves as a communication channel, where PLC is a server and PC is a client. OPC UA is a safe and independent platform for data exchange between PC and PLC. Control of the pneumatic valves and data collection from all sensors is defined by the mentioned data exchange which brings a high level of automation. PLC implements regulation algorithms and also performs checks of control parameters. The AMSV system uses intelligent image processing (in recognizing indicated dispenser data and in determining the volume of liquid in standard measuring containers). Image processing is performed using cameras that are connected to the PC via a POE switch. A sensor, such as a barcode reader (used for measuring instrument identification from measuring instrument database), environment sensors (temperature, pressure, and humidity), or a mass flow meter, communicates with PLC via TCP/IP protocol [15] through the switch [16]. All used valves have pneumatic propulsion with electric control and use a compressed air tank, in which air pressure is constantly measured.

An electro-pneumatic positioner [17] (used for setting flow during LPG measuring) is controlled by a (4–20) mA current, while START/STOP valves are controlled by voltage with the value of 24 V. Every START/STOP valve contains a pair of valve end position sensors with a relay output (connected as NO). These sensors are used for emptying standard measuring containers into collection tanks or when verifying LPD dispensers. Approximately 90% filling of collection tanks is indicated by the usage of level indicators. With temperature compensated dispensers, the temperature of liquids is measured in standard measuring containers. The biaxial inclinometer is used for the plane setting of standard measuring containers. A temperature and pressure meter with current output (4–20) mA is used for the verification of dispensers in the AMSV system.

The LTE Wi-Fi router creates a connection for the AMSV system (Figure 5) with the verifier's meter database via the internet with the usage of a GSM connection [18]. This connection can be used for uploading measurement data to the authentication provider´s cloud.

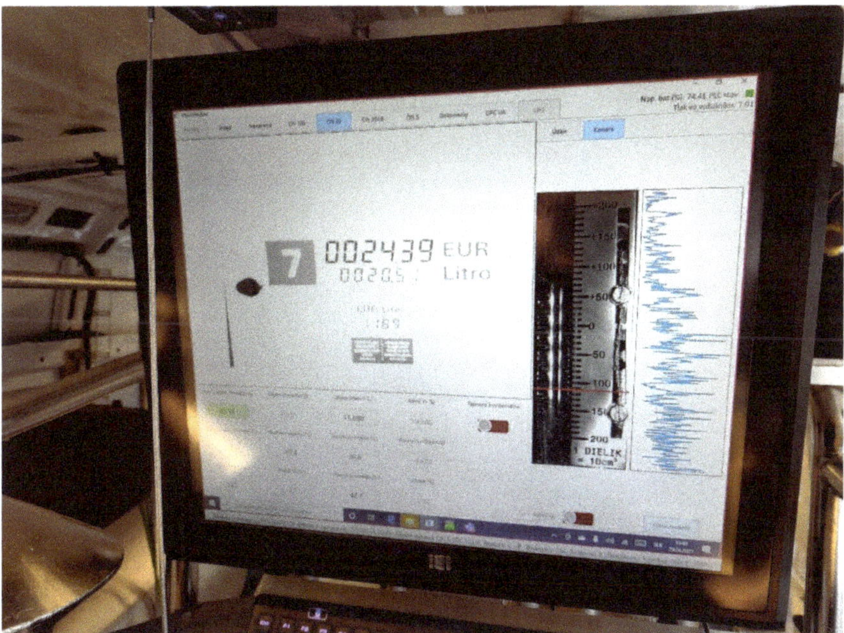

Figure 5. The AMSV system software with touch control.

4. Measurements

The AMSV system is a complex measuring device that is able to verify all available dispensers placed in a pump station thanks to its equipment. Measuring methods can be divided into two basic groups.

- Liquid volume measuring using standard measuring containers at atmospheric pressure (diesel, gas, AdBlue, and washer fluid).
- Liquid volume measuring using standard measuring containers in a closed circuit.

A high level of automation is used in every measurement.

Liquid volume in standard measuring containers as well as determination of indicated data of dispenser is realized through intelligent image processing.

Measurements that are not based on image processing are controlled by PLC (Figure 4). The following measurements directly affect liquid volume measurement:

- Biaxial tilt measurement of standard measuring containers.
- Temperature measuring of measured liquid (temperature compensated liquid volume).
- Measurement of environment temperature, pressure, and humidity.
- Measurement of the flowed liquid volume using a closed-loop mass flow meter (LPG).
- Closed circuit temperature measurement (LPG).
- Closed circuit pressure measurement (LPG).

Measurements which do not directly affect liquid volume measurement but are important in the process of automation:

- Indication of filled state of the collection tanks.
- Measuring the pressure in the compressed air tank: pneumatic valve factuality testing.
- Battery voltage measurement as a part of the power supply system.
- Monitoring the state of the end position of pneumatic valves.
- Meter barcode reading, for the unambiguous identification of meters in the verification provider´s database.

4.1. Liquid Volume Measurement Using Standard Measuring Tanks at Atmospheric Pressure

Measurement is performed by the value method with a stand-still start. The system offers four standard measuring containers with the goal to secure verification of the dispersion in the whole measuring range.

AMSV systems´ standard measuring containers are shown in Figure 6. The design of these standard measuring containers has a big advantage, which is minimal foam formation on the surface of the liquid during measurement. This shortens the waiting time for the right conditions which are needed to measure the received liquid value.

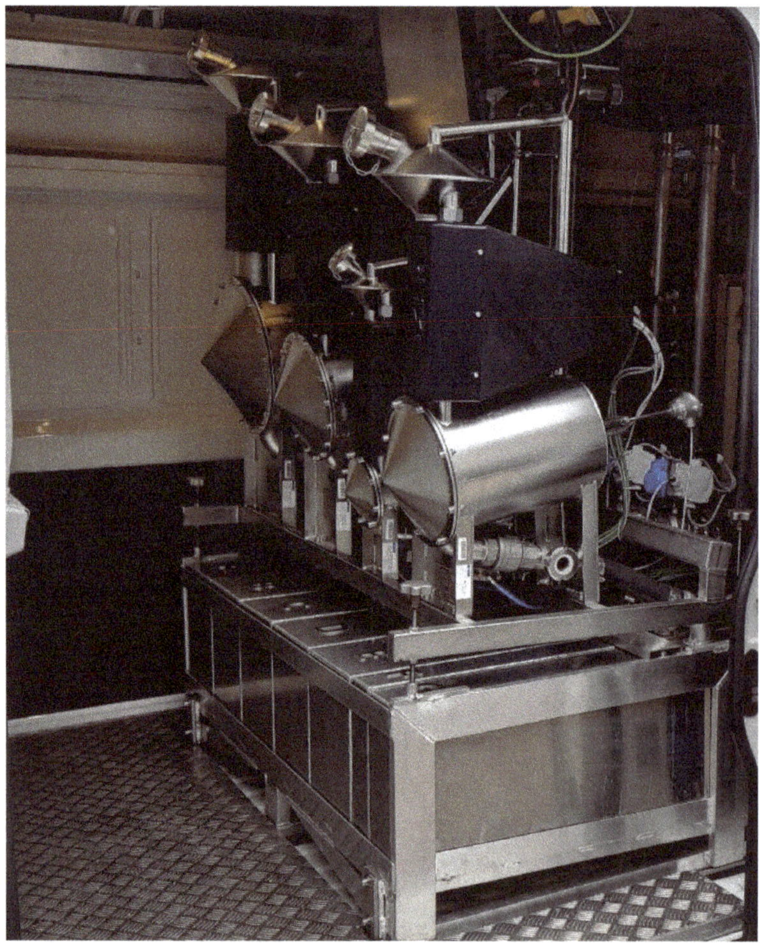

Figure 6. Standard measuring vessels used in the new generation of the AMSV system.

Intelligent image processing is used to measure the liquid volume in standard measuring containers (Figure 7).

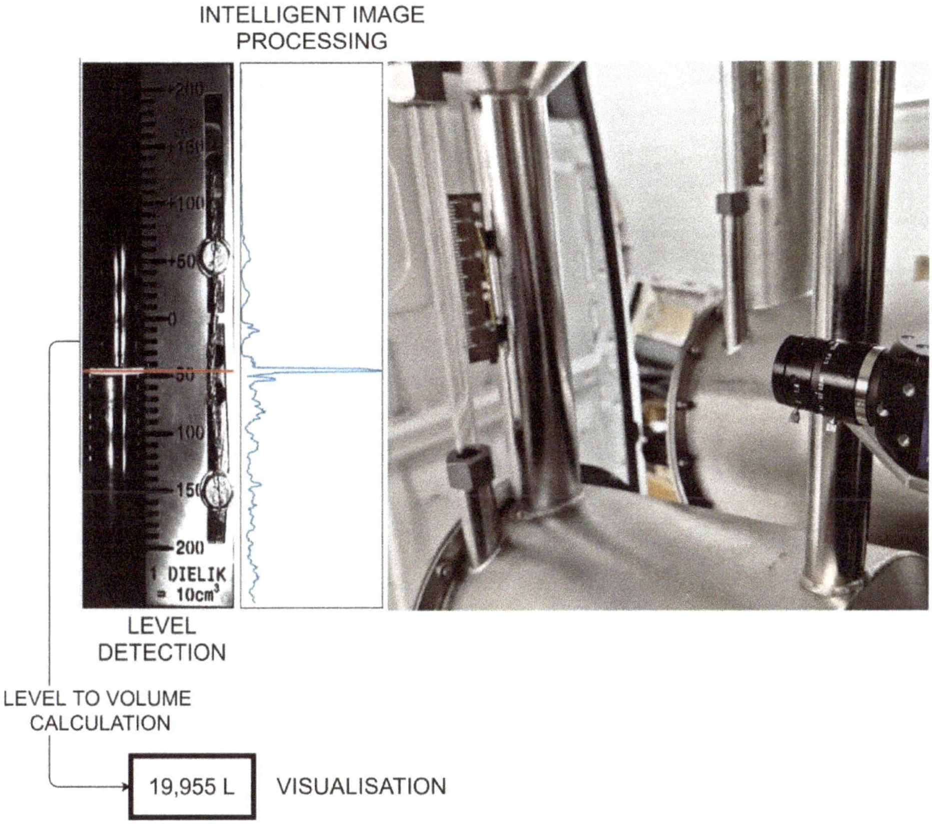

Figure 7. Determination of liquid volume in standard measuring vessels using intelligent image processing.

The camera captures the image of the status indicator, which is then input to the correlation-derivation algorithm. The algorithm evaluates the position of liquid level according to the image and performs a volume recalculation thanks to defined sensitivity. The issue of edge detection and machine vision is generally discussed in [19,20]. The volume is measured continuously in the range of the scale by this method. The value recognized by the algorithm is controlled by the metrological staff.

During the design process of liquid volume measurement, several measurement principles have been considered. The advantage of the chosen principle against other principles is significantly higher measuring dynamics. The radar level meter, for example, needs a certain time to stabilize the indicated value and its electronics are more sensitive to the temperature influence. Visual control of the measurement results is another big advantage of the optical principle (camera) for measuring liquid volume.

Intelligent image processing is also used to read indicated data from the dispenser display. Results digitalization by optical sign recognition, as part of automation, is an appropriate way due to legislation that determines the indicated value shown at the dispenser display as authoritative for payment. Intelligent image processing is the future trend in the field of legal metrology. Image recognition takes place in difficult conditions, which are formed from three factors:

- Dispensers at pump stations are manufactured by different companies with diverse types of indication of the amount of liquid dispensed.

- Surrounding light conditions: when placing a dispenser in the exterior, a wide range of light conditions is formed.
- Dispensers´ displays use plastic or glass protective cover which is the source of light reflection.

Reliable image processing can be in most cases impossible due to reflections created on the protective covers of displays. A POLARSENS chip-equipped camera can filter images in four polarization channels: $0°$, $45°$, and $135°$, resulting in reflection reduction. In the case when some indicated units indicate data in certain polarization, it is possible to choose a different polarization, which does not suppress indicated data of the indication unit. When the reflection causes inaccuracies, the measurement can be repeated.

An image without reflection is then used as input for the OCR algorithm. The OCR algorithm uses elements of artificial intelligence, in particular convolutional neural networks with deep learning [8]. The output is not only a recognized value, but also the probability with which this value was recognized. In this automated process, the technical worker just controls the recognized value.

4.2. Liquid Volume Measurement Using Standard Closedloop Mass Flow Meter (LPG Measurement)

Measurement is realized by a mass flow meter [21–24] (Figure 8) using the volume method with a standstill start (Figure 9).

The process of measurement is fully automated once the AMSV system is connected to the LPG dispenser. The technical staff sets required flows and volumes (Figure 10 shows an example of software part on which LPG verification is based) necessary to perform measurements.

The verification process is autonomously controlled by PC and PLC in the following steps:

The regulation process of required fluid flow was achieved by an electro–pneumatic positioner and controlled by a regulation algorithm implemented in PLC [25,26].

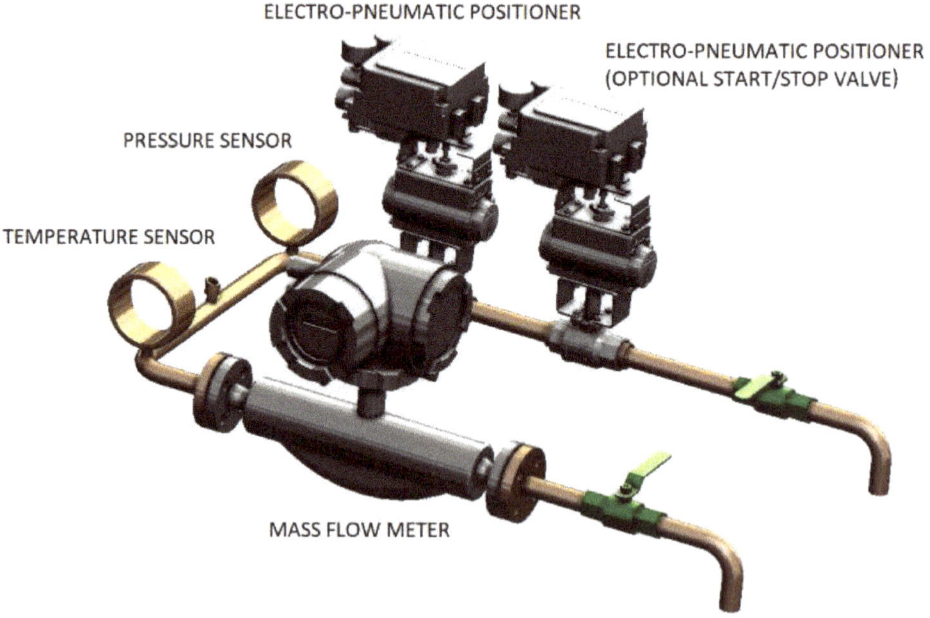

Figure 8. Standard measuring set for measuring the overflow volume of LPG.

Figure 9. Software user interface when verifying the LPG dispenser.

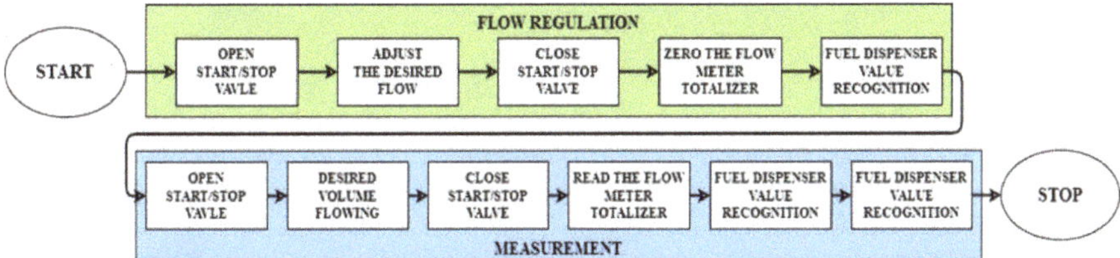

Figure 10. Fully automated LPG verification process.

5. Credibility of the Measurement Results

The credibility of the measurement results is an important factor.
The AMSV system uses three levels of results credibility control:

- Intelligent data processing: technical staff is not the only one who reads the data from the meters.
- In the case of changes in data recognition by technical staff, this change is recorded.
- Creating a so-called witness photo (Figure 11), where all key information of measurement is captured: dispenser indicated data, a level indicator of the standard measuring container, compensated values, time and place of measurement.

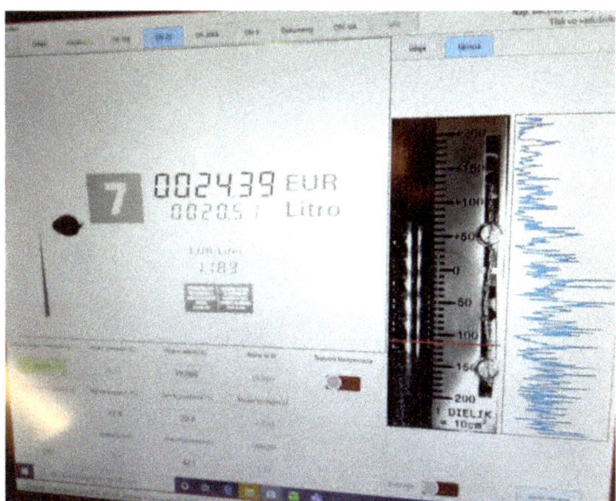

Figure 11. Witness photography as a check of the reliability of measurement results.

A witness photo of the measurement is shown in Figure 11.

6. Technical Specification of AMSV System

The AMSV system allows for the verification of dispensers available as standard at pump stations with the maximum flow to 200 L/min [27]. These are dispensers for:

- Standard hydrocarbon fuels (without temperature compensation).
- Standard hydrocarbon fuels (with temperature compensation).
- Liquefied natural gas (LPG).
- Washer liquid.
- AdBlue/uric acid.

The outcome of metrological control is composed of the following documents:

- Verification certificate.
- Calibration certificate.
- Confirmation of the liquid volume used during metrological control.
- Confirmation of performance.

The AMSV system has four standard containers, the characteristics of which are listed in Table 1.

Table 1. Metrological characteristics of AMSV system.

Standard Container	ON5	ON20AB	ON20	ON100
Specification	Washer liquid	AdBlue	Hydrocarbon fuels	Hydrocarbon fuels
Nominal volume (L)	5	20	20	100
Measuring range (L)	4.95–5.05	19.8–20.2	19.8–20.2	99–101
Volume of collecting tanks (L) and their number	90 (1×)	110 (1×)	240 (5×)	

The expanded measurement uncertainty complies with the requirements of OIML R 117-2 [28].

Once the volume measurement in the standard measuring container is completed, the liquid can be moved to a specific collection tank and the container can be reused. In the case of measuring hydrocarbon fuels, both standard measuring containers (ON20 and ON100) are connected to five collection tanks, which allows for the sorting of individual liquids (petrol diesel), as shown in Figure 12.

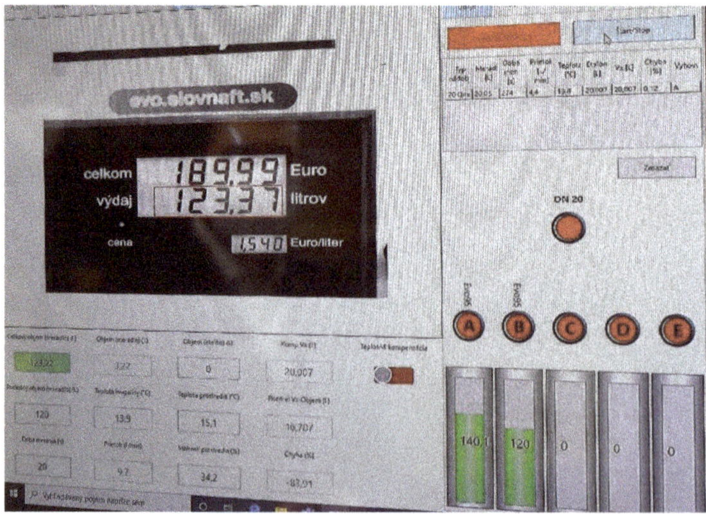

Figure 12. Method of sorting liquids into separate collection tanks (control software screenshot).

The collection tanks of hydrocarbon products can be drained using pumps with a flow rate of 50 L/min and the collection tanks for washer fluid and collection tanks for AdBlue can be drained with a flow rate of 30 L/min [22].

Other technical data are given in Table 2.

Table 2. Technical data of the mobile AMSV system.

AMSV			Characteristics
Batteries 24 V	External ~230 V	Recharging YES	Power supply
Operating *		Charging	Temperatures
−20 °C ~ +60 °C (outside temperature)		−10 °C ~ +40 °C (cabin temperature)	
	Volume		
240 L	110 L	90 L	
	Number		
5	1	1	Collection tanks
	Pump		
50 L/min	30 L/min	30 L/min	
Approximate time of depletion of the entire tank			
5 min	4 min	3 min	
GSM, Wi-Fi			Connectivity

* Temperature defines trouble-free usage of equipment, but the temperature at which metrological control can be performed can be modified by a special regulation according to the valid legislation.

7. Conclusions

Dispensers are one of the most used types of measuring instruments in business relationships worldwide. This fact was a motivation for the development of a new system generation for the metrological control of dispensers used at pump stations (AMSV system). The goal was to develop and manufacture a measuring system that would allow for realizing the verification of dispensers used at pump stations to increase credibility and efficiency, i.e., minimizing the downtime of dispensers as a source of financial loss of the seller. As mentioned in the article, the AMSV system uses automation in the form of intelligent image processing, complex data collection, monitoring of accompanying quantities, and their evaluation for completing the established goal.

The system has undergone complex tests and is currently used in the Slovak Republic.

Author Contributions: Conceptualization, J.M. and J.Ž.; methodology, J.M. and J.Ž.; investigation, J.M. and M.S.; data analysis, J.M. and M.S.; writing—original draft preparation, J.M. and P.T.; writing—review and editing, P.T. All authors have read and agreed to the published version of the manuscript.

Funding: The article was written in the framework of Grant Projects: VEGA 1/0318/21 "Research and development of innovations for more efficient utilization of renewable energy sources and for reduction of the carbon footprint of vehicles", KEGA 006TUKE-4/2020 "Implementation of Knowledge from Research Focused on Reduction of Motor Vehicle Emissions into the Educational Process".

Institutional Review Board Statement: Not applicable.

Informed Consent Statement: Not applicable.

Data Availability Statement: Not applicable.

Conflicts of Interest: The authors declare no conflict of interest.

References

1. Oiml.nim.ac.cn. Available online: https://oiml.nim.ac.cn/sites/default/files/media/document/2020-07/r117-1-e19.pdf (accessed on 19 January 2022).
2. Oiml.org. 2022. Available online: https://www.oiml.org/en/files/pdf_r/r117-1-e07.pdf (accessed on 19 January 2022).
3. Eur-lex.europa.eu. Available online: https://eur-lex.europa.eu/legal-content/EN/TXT/?uri=CELEX%3A32014L0032 (accessed on 20 January 2022).
4. Zpvvo.sk. 2022. Available online: https://www.zpvvo.sk/oiml-bulletin-july-2021_p40-48120648 (accessed on 20 January 2022).
5. Burdekin, F. General principles of the use of safety factors in design and assessment. *Eng. Fail. Anal.* **2007**, *14*, 420–433. [CrossRef]
6. Tools, D.; Architecture, U.; Concepts, P. OPC Factory Server. OPC Foundation. 2022. Available online: https://opcfoundation.org/developer-tools/specifications-unified-architecture/part-1-overview-and-concepts/ (accessed on 20 January 2022).
7. Davies, E.R. *Computer and Machine Vision*, 4th ed.; Academic Press: Cambridge, MA, USA, 2012.
8. Heaton, J.; Goodfellow, I.; Bengio, Y.; Courville, A. Deep learning. *Genet. Program. Evolvable Mach.* **2017**, *19*, 305–307. [CrossRef]
9. Plazas-Rosas, R.; Orozco-Gutierrez, M.; Spagnuolo, G.; Franco-Mejía, É.; Petrone, G. DC-Link Capacitor Diagnosis in a Single-Phase Grid-Connected PV System. *Energies* **2021**, *14*, 6754. [CrossRef]
10. Maniram Kumar, A.; Rajakarunakaran, S.; Pitchipoo, P.; Vimalesan, R. Fuzzy based risk prioritisation in an auto LPG dispensing station. *Saf. Sci.* **2018**, *101*, 231–247. [CrossRef]
11. Zubair, N.; Ayub, A.; Yoo, H. and Ahmed, I. PEM: Remote forensic acquisition of PLC memory in industrial control systems. *Forensic Sci. Int. Digit. Investig.* **2022**, *40*, 301336. [CrossRef]
12. Cheong, K.; Furuichi, N.; Doihara, R.; Kamazawa, S.; Kasai, S.; Hosobuchi, N. A comparison between a Coriolis meter and a combination method of a volumetric positive-displacement flowmeter and a densitometer in measuring liquid fuel mass flow at low flow rates. *Meas. Sens.* **2021**, *18*, 100321. [CrossRef]
13. Imam, N.; Vassilakis, V.; Kolovos, D. OCR post-correction for detecting adversarial text images. *J. Inf. Secur. Appl.* **2022**, *66*, 103170. [CrossRef]
14. Pástor, M.; Živčák, J.; Puškár, M.; Lengvarský, P.; Klačková, I. Application of Advanced Measuring Methods for Identification of Stresses and Deformations of Automotive Structures. *Appl. Sci.* **2020**, *10*, 7510. [CrossRef]
15. Brestovič, T.; Jasminská, N.; Čarnogurská, M.; Puškár, M.; Kelemen, M.; Fiľo, M. Measuring of thermal characteristics for Peltier thermopile using calorimetric method. *Measurement* **2014**, *53*, 40–48. [CrossRef]
16. Puškár, M.; Bigoš, P. Measuring of accoustic wave influences generated at various configurations of racing engine inlet and exhaust system on brake mean effective pressure. *Measurement* **2013**, *46*, 3389–3400. [CrossRef]
17. Puškár, M.; Jahnátek, A.; Kádárová, J.; Šoltésová, M.; Kovanič, Ľ.; Krivosudská, J. Environmental study focused on the suitability of vehicle certifications using the new European driving cycle (NEDC) with regard to the affair "dieselgate" and the risks of NOx emissions in urban destinations. *Air Qual. Atmos. Health* **2018**, *12*, 251–257. [CrossRef]
18. Toth, T.; Hudak, R.; Zivcak, J. Dimensional verification and quality control of implants produced by additive manufacturing. *Qual. Innov. Prosper.* **2015**, *19*, 9–21. [CrossRef]
19. Sabol, F.; Vasilenko, T.; Novotný, M.; Tomori, Z.; Bobrov, N.; Živčák, J.; Hudák, R.; Gál, P. Intradermal Running Suture versus 3M™ Vetbond™ Tissue Adhesive for Wound Closure in Rodents: A Biomechanical and Histological Study. *Eur. Surg. Res.* **2010**, *45*, 321–326. [CrossRef] [PubMed]
20. Sabol, R.; Klein, P.; Ryba, T.; Hvizdos, L.; Varga, R.; Rovnak, M.; Sulla, I.; Mudronova, D.; Galik, J.; Polacek, I.; et al. Novel Applications of Bistable Magnetic Microwires. *Acta Phys. Pol. A* **2017**, *131*, 1150–1152. [CrossRef]
21. Urban, R.; Štroner, M.; Kuric, I. The use of onboard UAV GNSS navigation data for area and volume calculation. *Acta Montan. Slovaca* **2020**, *25*, 361–374.
22. Pavlenko, I.; Saga, M.; Kuric, I.; Kotliar, A.; Basova, Y.; Trojanowska, J.; Ivanov, V. Parameter Identification of Cutting Forces in Crankshaft Grinding Using Artificial Neural Networks. *Materials* **2020**, *13*, 5357. [CrossRef] [PubMed]
23. Kuric, I.; Klačková, I.; Nikitin, Y.; Zajačko, I.; Císar, M.; Tucki, K. Analysis of Diagnostic Methods and Energy of Production Systems Drives. *Processes* **2021**, *9*, 843. [CrossRef]
24. Piňosová, M.; Andrejiová, M.; Lumnitzer, E. Synergistic effect of risk factors and work environmental quality. *Qual.-Access Success* **2018**, *19*, 154–159.
25. Puškár, M.; Markovič, J.; Král, Š.; Virgala, I.; Kopas, M.; Jančošek, M. Influence of biofuels on production of gaseous emission from diesel engine with regard to air quality. *Air Qual. Atmos. Health* **2020**, *13*, 763–772.
26. Lumnitzer, E.; Liptai, P.; Drahos, R. Measurement and Assessment of Pulsed Magnetic Fields in the Working Environment. In Proceedings of the 8th International Scientific Symposium on Electrical Power Engineering (Elektroenergetika), Stara Lesna, Slovakia, 16–18 September 2015; pp. 331–333.
27. Sinay, J.; Brestovič, T.; Markovič, J.; Glatz, J.; Gorzás, M.; Vargová, M. Analysis of the Risks of Hydrogen Leakage from Hydrogen-Powered Cars and Their Possible Impact on Automotive Market Share Increase. *Appl. Sci.* **2020**, *10*, 4292. [CrossRef]
28. Sopkuliak, P.; Palenčár, R.; Palenčár, J.; Suroviak, E.; Markovič, J. Evaluation of Uncertainties of ITS-90 by Monte Carlo Method. In *Computer Science Online Conference*; Springer: Cham, Switzerland, 2017; pp. 46–56. Available online: https://link.springer.com/chapter/10.1007/978-3-319-57264-2_5 (accessed on 1 April 2022).

Journal of
*Marine Science
and Engineering*

An Original Vibrodiagnostic Device to Control Linear Rolling Conveyor Reliability

Radka Jírová [1], Lubomír Pešík [1] and Robert Grega [2,*]

[1]   Faculty of Mechanical Engineering, Technical University of Liberec, Studenska 2,
      461 17 Liberec, Czech Republic; radka.jirova@tul.cz (R.J.); lubomir.pesik@tul.cz (L.P.)
[2]   Faculty of Mechanical Engineering, Technical University of Košice, Letna 9, 042 00 Kosice, Slovakia
*   Correspondence: robert.grega@tuke.sk; Tel.: +421-556022524

Abstract: On the basis of an analysis of the number of goods that are transported and handled in maritime transport, the ports for cargo ships may be considered as places with concentrated emissions. Reducing the emissions in ports can be achieved by shortening the stay times of cargo ships. The time that ships spend in ports may be reduced to the time that is required for the effective handling of the goods. One of the solutions for effective handling is using equipment with linear rolling systems. To prevent the idle time of cargo ships and the unnecessary increment of emissions in ports because of the possible failure of the linear rolling systems, their reliability and failure prediction are greatly required. Unfortunately, the common diagnostic systems of linear rolling systems in transportation practice still fail in particular cases of great external loads. Therefore, an original solution of the diagnostic system was designed on the basis of a load-free diagnostic part with a vibration sensor that is integrated into a carriage of the linear rolling system. A functional sample of the diagnostics was produced, and the vibrations that were measured on a loaded carriage and on the diagnostic part were compared in laboratory conditions under significant external loads. Encouraging results were reached by a time-domain analysis of the measured data. On the diagnostic part, the damage appeared clearly, while, on the loaded carriage, there were no observable signs of damage.

Keywords: linear rolling systems; ship emissions; vibrodiagnostics; wear

Citation: Jírová, R.; Pešík, L.; Grega, R. An Original Vibrodiagnostic Device to Control Linear Rolling Conveyor Reliability. *J. Mar. Sci. Eng.* **2022**, *10*, 445. https://doi.org/10.3390/jmse10030445

Academic Editor: María Isabel Lamas Galdo

Received: 27 January 2022
Accepted: 19 March 2022
Published: 21 March 2022

Publisher's Note: MDPI stays neutral with regard to jurisdictional claims in published maps and institutional affiliations.

1. Introduction

Because of significant climate change, which is increasing GHG (greenhouse gas) emissions and global warming, the sustainability of maritime transport is currently in demand. As maritime transport creates 80% of the global transportation of goods, the necessity of decreasing the emissions that are related to container and other cargo ships is obvious. The United Nations [1,2] mentions that, for the effective reduction in the GHG emissions in maritime transport, improvements in the port logistics should be included.

Ports can optimise and ensure the availability of the berth at the moment when ships arrive in order to avoid idle time that leads to an unnecessary increment of the emissions in ports [3–6]. Logistics problems by berthing containers and cargo ships may be caused by insufficient communication between the ports and the ships, as well as by internal logistics challenges that are connected with the unloading/loading of goods, or with the transportation of goods within the port [7–10].

Especially in the case of internal transportation, the manipulation and positioning of containers and goods, which are requirements in logistics operations, are increasing so that the processes will be fast, effective, and reliable. To secure this assumption, increased attention must be paid to the reliability and failure prediction of the crane subcomponents and their positioning, and of the clamping systems and handling machines. In transportation practice, linear rolling systems that realise the linear motion of these machines and that are the basis of their design, are said to be critical [11–15]. Their failure may lead to the breakdown of the machine, the collapse of the logistic chain, and, thus, to an increase in the emissions that are produced by the ships that are waiting for the berth. Therefore, special emphasis is given to the diagnostics and failure prediction of the linear rolling systems that are used in port logistics.

The diagnostics of linear rolling systems is based on the knowledge of rolling bearing diagnostics. In the diagnostics of rolling bearings, the methods usually aim to measure the vibrations or the acoustic emissions. However, the transition from bearing diagnostics to the diagnostics of linear rolling systems is not so simple. Japanese producer, THK (THK Co., Ltd., Tokyo, Japan), in the patent [16], mention that the design of linear systems might negatively influence the diagnostics on the basis of the measurement of the vibrations or the noise intensity. The transition of the rolling elements from a nonloaded state to a loaded one excites vibrations that are similar to those created by the damage of the guiding profile in the time domain. In the frequency domain, the signal that belongs to the changed status shows a frequency that is identical to the guiding profile damage.

Therefore, researchers and the producers of linear rolling systems developed diagnostic systems that are based on measuring the acceleration of the vibrations, and on interpreting the measured vibrations through the RMS (root mean square) values, a spectral analysis, and the crest factor [17], or a spectrogram [18,19], in the context of the lubrication level, which provides the reference value of the RMS. Feng [20] explains the influence of lubrication on the RMS and its distribution in the frequency domain. He detected an increased RMS in a high-frequency band by progressing damage.

The producer of THK, in its patent [21], argues for the use of an acceleration sensor for measuring the vibrations, and for the use of a presence sensor of the rolling elements to determine the actual velocity of the linear system. The vibrations are evaluated through the RMS values in the high-frequency band, which reflect the lubrication level and the linear system velocity. When the experimentally determined limit value is exceeded, the linear system is relubricated, and the level of the vibrations is measured again.

The producer, Schaeffler (Schaeffler Technologies AG & Co. KG, Herzogenaurach, Germany), brought in a similar method of diagnostics [22] that uses the acceleration sensor for measuring vibrations. The measured vibrations are evaluated through the RMS values in the high-frequency band from 14 kHz to 25 kHz. When the limit value is exceeded, the linear system is relubricated. A decisive quantity is the time of a relubricating process compared to the experimentally given distribution function of the linear system life. The analysing procedure is processed through neural networks.

Unfortunately, the diagnostics of linear rolling systems that are currently provided by producers still do not appear reliable enough in transportation practice. In practice, several cases of significant damage to the guiding profiles (Figure 1) and the rolling elements (Figure 2) may be noticed. In these cases, the linear rolling systems were operated under significant external loads by the weight of the transported goods. The measured vibrations did not exceed the threshold value, and the linear systems were considered serviceable according to the vibrodiagnostics that was used. Thus, the paper aims to describe the development of an original principle for linear rolling system diagnostics.

Figure 1. Damage of the guiding profile of linear rolling system.

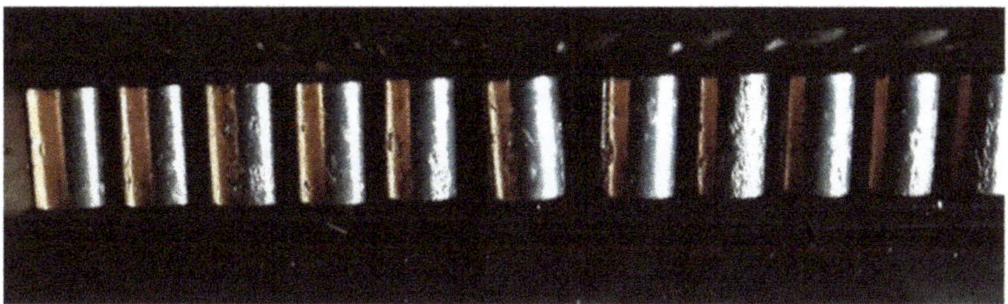

Figure 2. Pitting of rolling elements of linear rolling system.

2. Materials and Methods

In the specific case of handling machines, linear rolling systems are frequently operated under enormous loads. These loads are related to the mass inertia of the connected mechanical assemblies and of the transported goods. The mass inertia causes a dynamical load that is composed of static and dynamical forces. In this context, it may be noticed that the dynamical character of the load negatively affects the life of the linear system [23]. Knowledge of the actual load is also beneficial for the diagnostic function. It might be recognised that the external dynamical load influences the measured vibrations by way of damping them rapidly, or exceeding them in orders, by the dynamical behaviour of the machine.

Thus, an original principle of the linear rolling system with an integrated diagnostic system was proposed [24]. A load-free part was designed in the linear system carriage, with shared rolling elements. Through the load-free part, the external dynamical load is eliminated, and the vibrations that are measured on the diagnostic (load-free) part more easily enable the recognition of the damage (Figure 3).

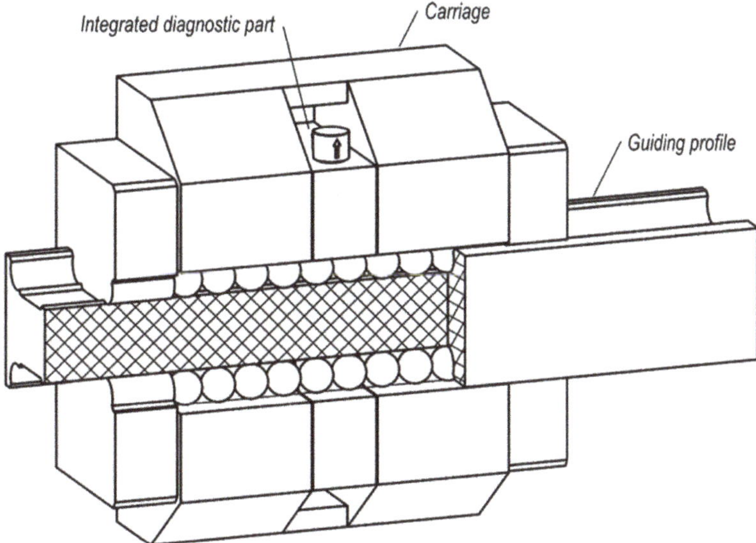

Figure 3. The linear system carriage with the integrated diagnostic part.

For the verification of the proposed principle, a functional sample was prepared and tested. The testing process, which was divided into two stages, was based on a simulation of the operating conditions and the guiding profile damage. The first stage compared the vibrations that were measured on a loaded and a load-free carriage in three directions. The second stage focused on evaluating the vibrations that were measured on the diagnostic part of the functional sample. Both measurements were processed in the time domain.

The constructed testing facility enabled the pressure loading of the linear rolling systems by pneumatic springs. A linear motion was ensured by an electromotor through a rope drive (Figures 4 and 5). When testing, the guiding profile assembly moved by velocity ($v = 0.42$ ms^{-1}), while the tested carriages and the functional sample stood still and were connected by pneumatic springs to the frame. The pressure load that was used for the testing was $F = 18.5$ kN.

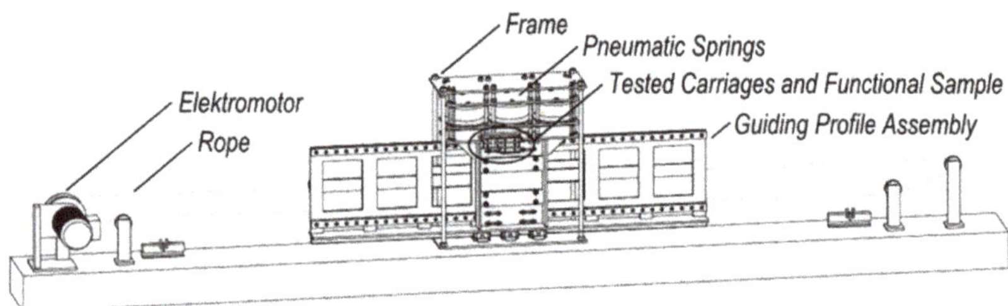

Figure 4. The model of the testing facility.

Figure 5. The testing facility and vibrations measurement.

For the functional sample, which was based on the diagnostic part (Figure 6), the Hiwin (Hiwin Technologies Corp., Taichung, Taiwan) linear rolling system was redesigned. Its basic parameters are summarised in Table 1. An iron unit of the carriage was divided into three pieces, with an appropriate clearance fit between them, by combining tinny spacers. The final clearance fit was in the order of hundredths.

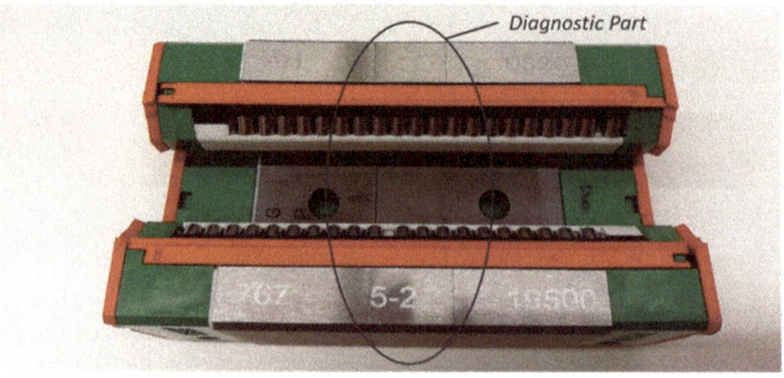

Figure 6. The functional sample.

Table 1. Parameters of Hiwin linear rolling system.

Type	Basic Dynamic Capacity	Rolling Elements	Diameter of Rolling Element
Hiwin RGH30CA ZAH	39.1 kN	Rollers	$d_v = 4$ mm

Figure 7 shows the guiding profile damage as a ground groove, with a thickness of 1 mm on the contact surface. The damage represented a fatigue failure that was caused by contact pressure, which occurs by the rolling motion.

Figure 7. The simulated damage of the guiding profile.

In the first stage of testing, the acceleration sensors were placed according to the scheme in Figure 8. Three-axis sensors were applied for evaluating the vibrations in all three axes. The three-axis sensor of MMF (type: KS903.100) (Manfred Weber Metra Meß- und Frequenztechnik in Radebeul e.K., Radebeul, Germany) featured a range of ±60 G, and a linear frequency range up to 10 kHz. With respect to the linear frequency range of the sensors, a sampling frequency of the measurement was set to 20 kHz. Figure 9 shows the actual sensor positions on the loaded and the load-free carriage.

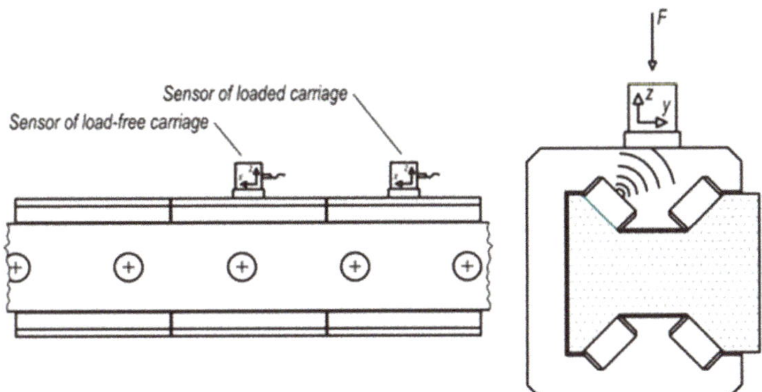

Figure 8. The scheme of sensor position against the damage.

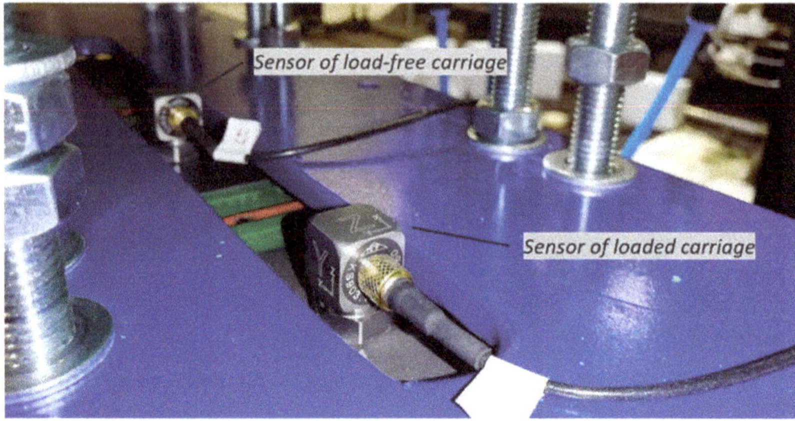

Figure 9. The placing of three-axis acceleration sensors.

A similar method to the first stage was employed for testing the functional sample with the integrated diagnostic part. The one-axis acceleration sensors were placed according to the scheme in Figure 10. The one-axis sensor of the MMF producer (type: KS97.100) featured the range of ±60 G and a linear frequency range up to 13 kHz. With consideration to the linear frequency range of the sensors, the sampling frequency was set to 25 kHz. Figure 11 shows the actual sensor positions on the loaded carriage and the diagnostic part of the functional sample.

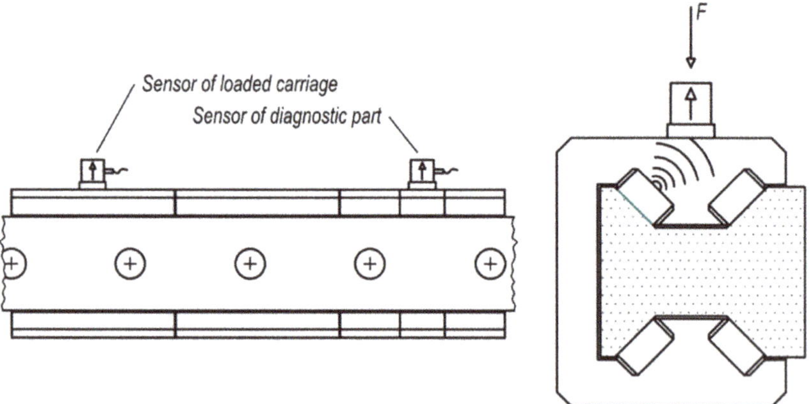

Figure 10. The scheme of sensor position against the damage.

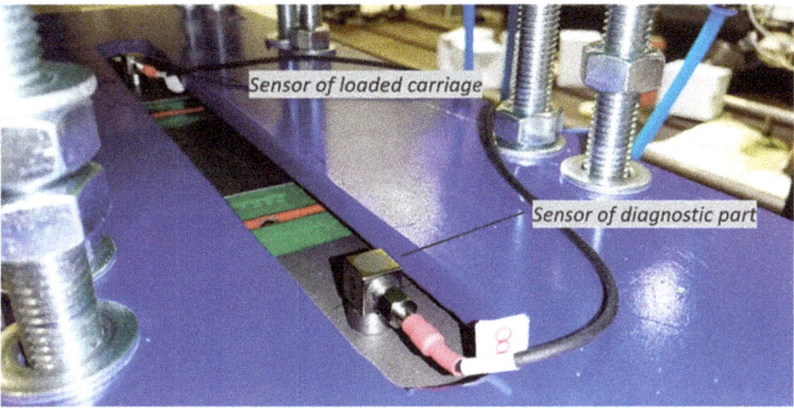

Figure 11. The placing of one-axis acceleration sensors.

3. Theory

In the service of linear rolling systems, the contact pressure between the guiding profile and the rolling elements appears because of the external dynamical loads. The cyclic loading and unloading of linear systems causes a relative shift in the contact surfaces that is related to the different radii of their curvatures. Furthermore, the linear motion is characterised by the rolling process of the rolling elements against the guiding profile. The rolling motion leads to the asymmetric deformation of the contact surfaces and to the asymmetric distribution of the contact pressure. This behaviour belongs to the hysteresis of the used material [25,26].

Dynamical changes in the contact pressure as a result of the external load and the rolling motion may generate a fatigue failure, which is called "pitting" or "spalling". The fatigue failure is represented by a breakout of the contact surface particles, which leaves pits at sufficient lubrication, or spalling at the surface layer at insufficient lubrication [27,28].

The fatigue failure of the rolling elements and the guiding profiles produces a rolling friction increment that decreases the linear system efficiency [29–32]. The result of the fatigue failure is increases in the temperature, the noise, and the vibrations of linear systems. Vibrations may negatively affect related bodies and even the linear system itself, which contributes to their development.

Vibrations present increased amplitudes with regard to the frequency or the time period of the damage. Frequencies may be computed through kinetic analysis for two general cases: for the damage of the rolling element (Figure 12), or for the damage of the guiding profile (Figure 13). According to kinematic schemes, the damage frequencies and time periods depend on the linear motion velocity.

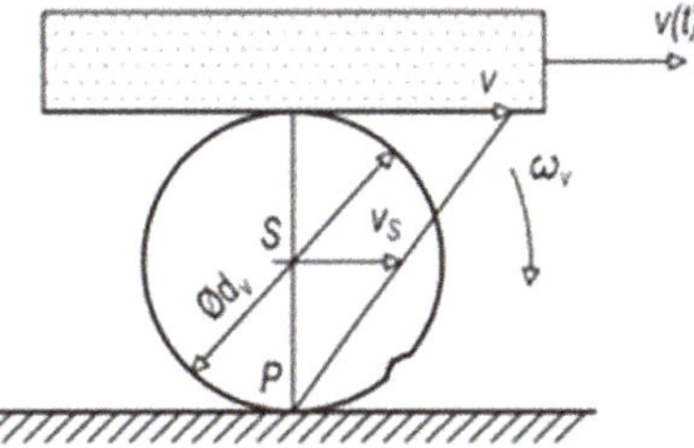

Figure 12. The kinematic scheme—the damage of the rolling element.

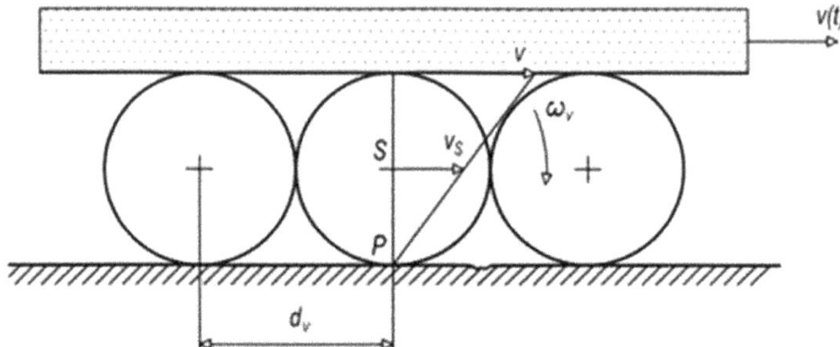

Figure 13. The kinematic scheme—the damage of the guiding profile.

In order to derive the damage frequency of the rolling element, it may be stated that its velocity at the point (P) equals zero, so that:

$$v_P(t) = v_S(t) - \omega_v(t)\frac{d_v}{2} = 0 \tag{1}$$

where the velocity of the rolling element at the rotation centre ($v_s(t)$) is related to the velocity of the linear motion ($v(t)$), by the equation:

$$v(t) = v_S(t) + \omega_v(t)\frac{d_v}{2} \tag{2}$$

Then, the velocity ($v_s(t)$) may be computed as:

$$v_S(t) = \frac{v(t)}{2} \tag{3}$$

Through the diameter of the rolling element (d_v), the equation for the angular velocity of the rolling element is:

$$\omega_v(t) = \frac{2v_S(t)}{d_v} \tag{4}$$

The damage time period of the rolling element reflects the vibration excitation by the damage twice in one rotation of the rolling element, at the upper and lower contact surfaces:

$$T_{Dv}(t) = \frac{2\pi}{2\omega_v(t)} = \frac{\pi d_v}{v(t)} \tag{5}$$

Then, the damage frequency is:

$$f_{Dv}(t) = \frac{1}{T_{Dv}(t)} = \frac{v(t)}{\pi d_v} \tag{6}$$

The damage time period of the guiding profile refers to the time when the rolling element performs a distance of d_v by the velocity ($v_S(t)$) of the rotation centre:

$$T_{Dp} = \frac{d_v}{v_S(t)} = \frac{2d_v}{v(t)} \tag{7}$$

The damage frequency equals:

$$f_{Dp}(t) = \frac{1}{T_{Dp}(t)} = \frac{v(t)}{2d_v} \tag{8}$$

For the case of the functional sample, the time periods and the frequencies of damage are summarised in Table 2.

Table 2. Damage time periods and frequencies of Hiwin linear rolling system.

Rolling Element		Guiding Profile	
$T_{Dv} = 0.03$ s	$f_{Dv} = 33.4$ Hz	$T_{Dp} = 0.019$ s	$f_{Dp} = 0.019$ Hz

With reference to the proposed principle of the linear system diagnostics on the basis of the integrated diagnostic part, the negative influence of the external loads was observed on the linear system through a simplified mechanical model. In the mechanical model, elastic and damping connections may substitute for the rolling elements [32–35], and the damage may be represented by kinematical excitation (Figure 14).

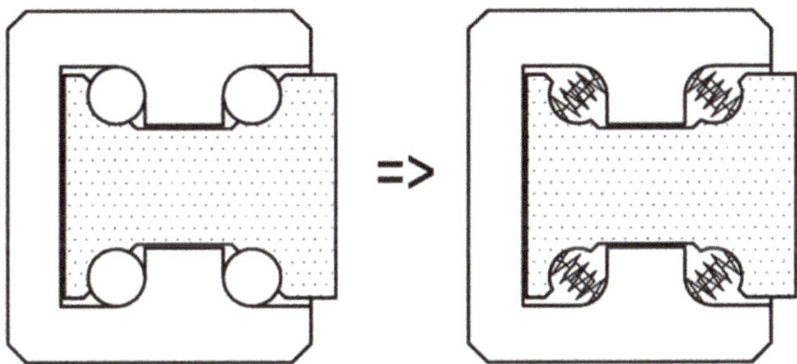

Figure 14. The substitution of rolling elements by elastic and damping connections.

For the kinematical excitation, a hyperbolic tangent function was applied to supplant a discontinuous function that was shaped similar to a rolling element crossing over the damage with a thickness of 1 mm. Figure 15 presents the kinematical excitation with the time period of the guiding profile damage.

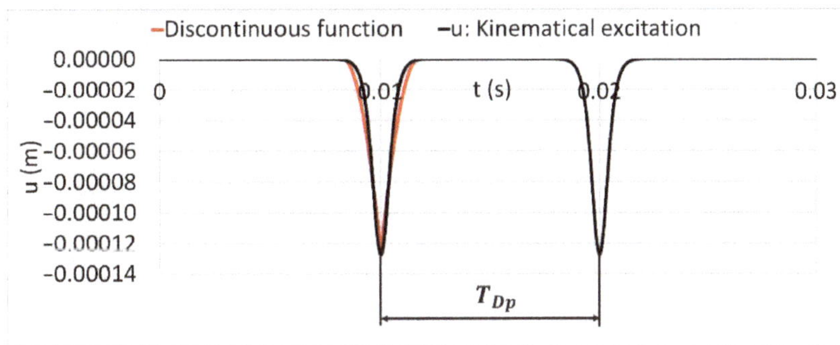

Figure 15. The kinematical excitation compared to the discontinuous function.

The simplified mechanical model (Figure 16) substitutes elastic connections through reduced stiffness in a radial direction that was provided by the producer's data. For the damping connections, we used a damping ratio with a value of $b_{rel} = 0.5$ [36]. A mass of m represents the mass of the carriage unit, and, in the loaded case, M signifies the mass of the carriage unit and the inertia mass of the hanged bodies.

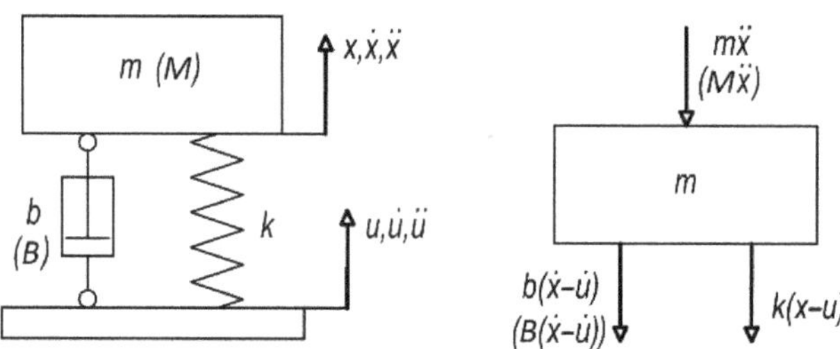

Figure 16. The simplified mechanical model.

Table 3 summarises the dynamical parameters of the simplified mechanical model.

Table 3. Dynamical parameters of the simplified mechanical model in Figure 16.

Mass of The Carriage Unit	Mass of The Carriage Unit and Hanged Bodies	Stiffness of Rolling Elements	Damping Coefficient of The System without Hanged Bodies (including Sealing Parts and Lubrication)	Damping Coefficient of The System with Hanged Bodies (including Sealing Parts and Lubrication)
$m = 0.82$ kg	$M = 100$ kg	$k = 1.2 \times 10^6$ N·m^{-1}	$b \cong 1 \times 10^3$ Ns·m^{-1}	$B \cong 1 \times 10^4$ Ns·m^{-1}

The result of the motion differential in Equation (9) is the time relation of the mass acceleration as a system response to the excitation function. For the case without hanged bodies to the carriage, the motion differential equation equals:

$$m\ddot{x} + b\left(\dot{x} - \dot{u}\right) + k(x - u) = 0 \tag{9}$$

With hanged bodies, the motion differential equation equals:

$$M\ddot{x} + B\left(\dot{x} - \dot{u}\right) + k(x - u) = 0 \tag{10}$$

where a damping coefficient (b) equals:

$$b = 2b_{rel}\sqrt{mk} \tag{11}$$

$$B = 2b_{rel}\sqrt{Mk} \tag{12}$$

The motion differential in Equations (9) and (10) were processed in MATLAB software, and the results are shown in Figure 17.

For testing the diagnostic function, the linear rolling system loaded by an external force may be represented through the simplified mechanical model in Figure 18. The external force as the load from the pneumatic springs acts similarly to the mass inertia. By the elastic deformation of the rolling elements, a preload of the linear rolling systems is reached. The preload leads to the system stiffness increase.

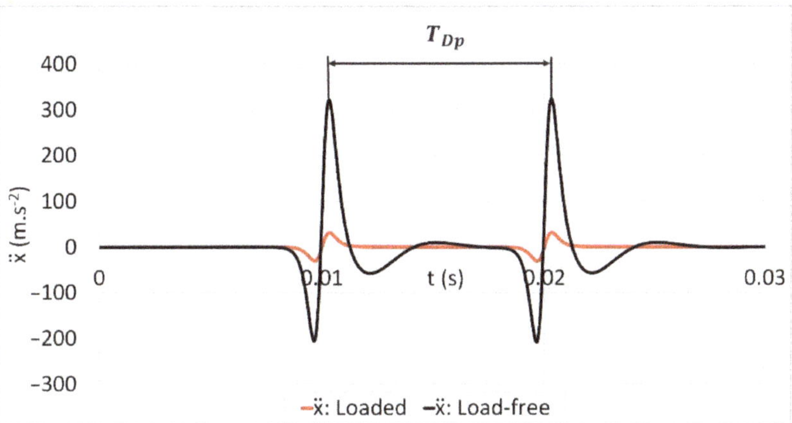

Figure 17. The time relation of mass acceleration—the influence of the mass inertia.

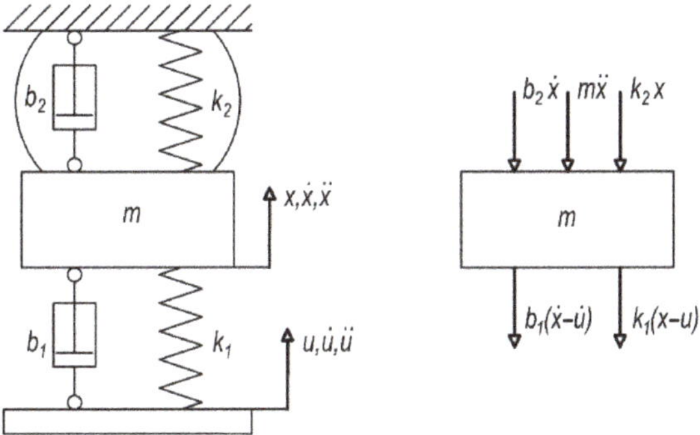

Figure 18. The simplified mechanical model loaded by the external force.

In the mechanical model, the stiffness (k_1) and the damping coefficient (b_1) are related to the rolling elements, while k_2 and b_2 belong to the pneumatic springs. The result of the motion differential in Equation (13) is the time relation of the mass acceleration as a system response to the excitation function:

$$m\ddot{x} + b_1\left(\dot{x} - \dot{u}\right) + b_2\dot{x} + k_1(x - u) + k_2x = 0 \tag{13}$$

Table 4 summarises the dynamical parameters of the simplified mechanical model loaded by the external force.

Table 4. Dynamical parameters of the simplified mechanical model in Figure 18.

Mass of The Carriage Unit	Stiffness of Preloaded Rolling Elements	Stiffness of Pneumatic Spring	Damping Coefficient of The System (including Sealing Parts and Lubrication)	Damping Coefficient of Pneumatic Springs
$m = 0.82$ kg	$k_1 = 3.6 \times 10^7$ N·m^{-1}	$k_2 = 8 \times 10^3$ N·m^{-1}	$b_1 \cong 5 \times 10^3$ Ns·m^{-1}	$b_2 \cong 80$ Ns·m^{-1}

The motion differential in Equation (13) was processed in MATLAB software, and the results are shown in Figure 19.

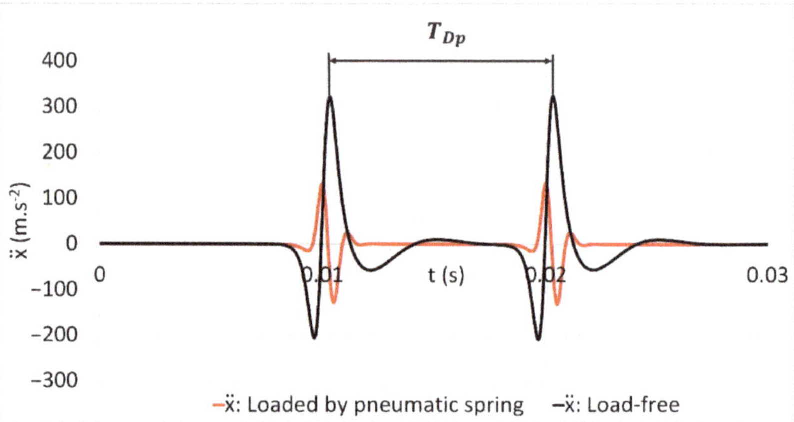

Figure 19. The time relation of mass acceleration—the influence of the external force.

In both cases, loading the linear rolling system by the external load in the form of mass inertia or external force showed a significant reduction in the acceleration ($\ddot{x}$), which was excited by the damage.

4. Results and Discussion

The diagnostic function of the proposed diagnostic principle was verified via the testing facility under the abovementioned conditions. On the contact surface of the guiding profile, the groove was ground for the damage simulation. Then, the vibrations of the loaded and the load-free carriage, and the diagnostic part of the functional sample, were measured and analysed in the time domain.

In the first stage of testing, the influence of the external load and the measurement direction was decided for vibrations that were excited by the simulated damage. The load-free carriage was placed between two that were tightly connected to the testing facility, and, finally, the mentioned effects were evaluated.

The time graph in Figure 20 shows the measured acceleration of the vibrations in the time domain at the loaded carriage crossing over the simulated damage. Major amplitudes of the measured acceleration may be recognised at a time period of T_{cBA} with the parameters summarised in Table 5.

Table 5. Time period and its parameters according to Figure 20.

Parameters		Time Period
t_{cA} = 7.24420 s	t_{cB} = 7.26425 s	T_{cBA} = 0.02 s

The time period (T_{cBA}) equals:

$$T_{cBA} = t_{cB} - t_{cA} \tag{14}$$

These amplitudes might be observed through the entire time of the carriage motion. Thus, it may be noted that the major amplitudes belong to the changed status of the rolling element from nonloaded to loaded. The amplitudes of acceleration that could be related to the simulated damage are hidden in the vibration noise.

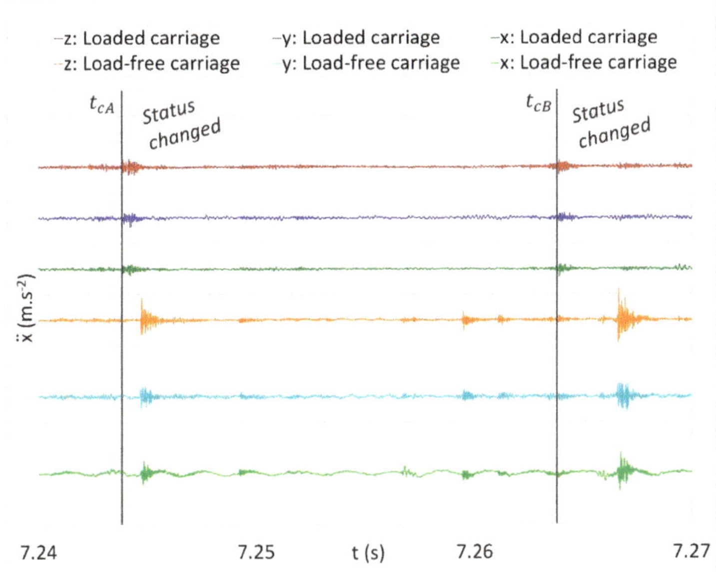

Figure 20. The time graph of acceleration by loaded carriage crossing over the simulated damage: ordinate tick marks: $100\ ms^{-1}$, processed in DeweSoft.

The time graph in Figure 21 illustrates the measured acceleration of the vibrations in the time domain at the load-free carriage crossing over the simulated damage. The time parameters of the major (T_{c42}) and minor (T_{c31}) acceleration amplitudes may be detected with the parameters that are specified in Table 6.

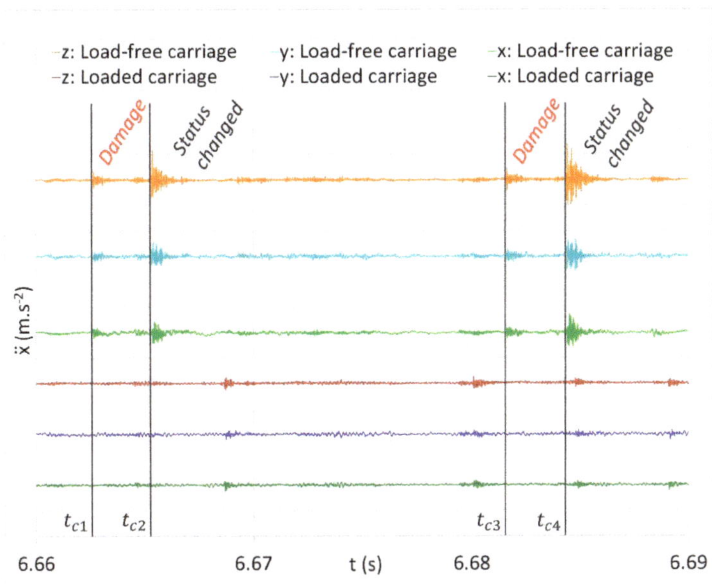

Figure 21. The time graph of acceleration by load-free carriage crossing over the simulated damage: ordinate tick marks: $100\ ms^{-1}$, processed in DeweSoft.

Table 6. Time periods and their parameters according to Figure 21.

Time Parameters				Time Periods	
$t_{c1} = 6.66275$ s	$t_{c2} = 6.66555$ s	$t_{c3} = 6.68135$ s	$t_{c4} = 6.68455$ s	$T_{c21} = 0.0023$ s	$T_{c43} = 0.0027$ s

As can be seen in the time graph, the major amplitudes of the acceleration with the time period of T_{c42} belong to the changed status of the rolling element from nonloaded to loaded. In the measured vibrations, minor amplitudes of acceleration (T_{c31}) may be observed. These amplitudes are related to the vibrations that were excited by the simulated damage:

$$T_{c42} = t_{c4} - t_{c2} \tag{15}$$

$$T_{c31} = t_{c3} - t_{c1} \tag{16}$$

The time period of the major and minor amplitudes is approximately equal to the damage time period of the guiding profile: $T_{c42} \cong T_{c31} \cong T_{Dp}$. Through the time phases, ΔT_{c21} and ΔT_{c43}, between the major and minor amplitudes, the damage position against the current position of the rolling elements can be determined:

$$\Delta T_{c21} = t_{c2} - t_{c1} \tag{17}$$

$$\Delta T_{c43} = t_{c4} - t_{c3} \tag{18}$$

The distance (Δl_c) of the simulated damage against the rolling elements is roughly equal to:

$$\Delta l_c \cong \frac{1}{2}\Delta T_{c21}v \cong \frac{1}{2}\Delta T_{c43}v \tag{19}$$

Through the length of the iron unit (l_c), a number of rolling elements (n_c) in contact with one raceway of the guiding profile may be calculated:

$$n_c = \frac{l_c}{d_v} \tag{20}$$

The results of Equations (17)–(20), which are summarised in Table 7, indicate 17 or 18 rolling elements in contact with the raceway at each time (Figure 22).

Table 7. Results of Equations (17)–(20).

Time Phases		Distance of Simulated Damage against Rolling Elements	Length of The Iron Unit	Number of Rolling Elements in Contact with One Raceway of Guiding Profile
$\Delta T_{c21} = 0.0023$ s	$\Delta T_{c43} = 0.0027$ s	$\Delta l_c \cong 0.5$ mm	$l_c = 71$ mm	$n_c = 17.75$

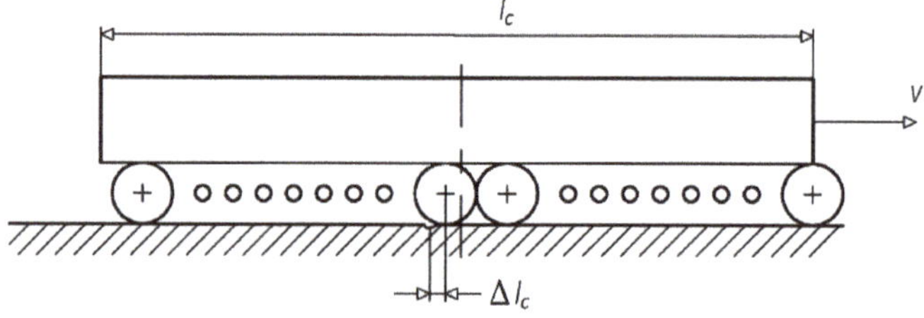

Figure 22. The position of the simulated damage against the rolling elements.

According to the analysis of the measured acceleration, it may be stated that the external load of linear systems is directly related to the early detection of possible failure. In the loaded carriage, the measured amplitudes of acceleration are significantly lower than in the load-free carriage; and the amplitudes related to the vibrations that were excited by the simulated damage are hidden in the vibration noise.

The acceleration of vibrations was measured and evaluated in three coordinate axes: x, y, and z. In the case of the loaded carriage, the amplitudes of acceleration in all the axes reached similar values. Therefore, vibrations can be measured in any direction, and their amplitudes are independent of the loading force direction. In the load-free carriage, a difference may be observed between the amplitudes measured in the z-axis (radial) direction and the amplitudes measured in the other two axes directions. This attribute might be related to the gravitational acceleration that acts in the same direction. Thus, the one-axis sensor was placed in the appropriate direction for the subsequent testing.

In the second stage, the functional sample with the integrated diagnostic part was analysed. The time graph in Figure 23 shows the measured acceleration of the vibrations in the time domain at the loaded carriage crossing over the simulated damage. Major amplitudes of measured acceleration may be recognised at a time period of T_{dBA} with the parameters that are summarised in Table 8:

$$T_{dBA} = t_{dB} - t_{dA} \tag{21}$$

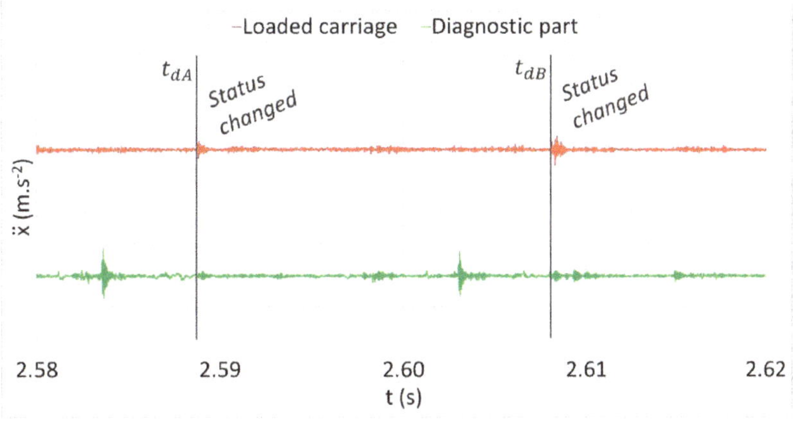

Figure 23. The time graph of acceleration by loaded carriage crossing over the simulated damage: ordinate tick marks: 100 ms^{-1}, processed in DeweSoft.

Table 8. Time period and its parameters according to Figure 23.

Time Parameters		Time Period
t_{dA} = 2.58904 s	t_{dB} = 2.60804 s	T_{dBA} = 0.019 s

It might be noted that these amplitudes belong to the changed status of the rolling element from nonloaded to loaded. In contrast, the amplitudes of acceleration related to the simulated damage are hidden in the vibration noise.

The time graph in Figure 24 illustrates the measured acceleration of the vibrations in the time domain at the diagnostic part crossing over the simulated damage. The time parameters of the major (T_{d42}) and minor (T_{d31}) acceleration amplitudes may be detected with the parameters that are specified in Table 9.

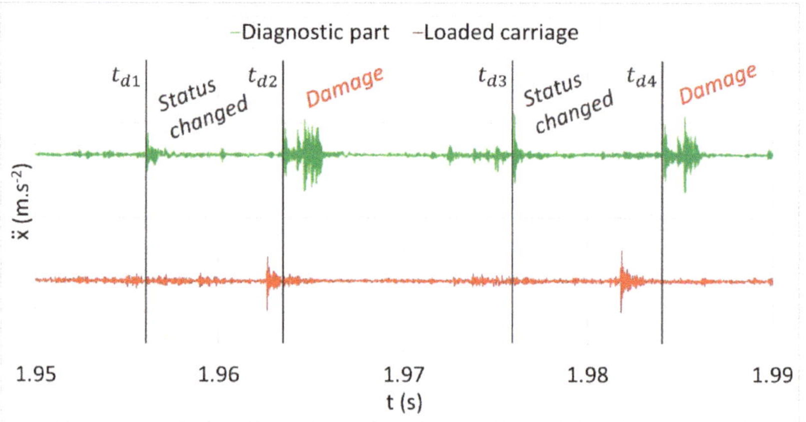

Figure 24. The time graph of acceleration by diagnostic part crossing over the simulated damage: ordinate tick marks: 100 ms^{-1}, processed in DeweSoft.

Table 9. Time periods and their parameters according to Figure 24.

Parameters				Time Periods	
t_{d1} = 1.95636 s	t_{d2} = 1.96376 s	t_{d3} = 1.97620 s	t_{d4} = 1.98432 s	T_{d21} = 0.0206 s	T_{d43} = 0.0198 s

The time periods, T_{d42} and T_{d31}, are equal to:

$$T_{d42} = t_{d4} - t_{d2} \tag{22}$$

$$T_{d31} = t_{d3} - t_{d1} \tag{23}$$

As can be seen in the time graph, the major amplitudes of acceleration with the time period of T_{d42} belong to the vibrations that were excited by the simulated damage. In the measured vibrations, minor amplitudes of acceleration (T_{d31}) may be observed. These amplitudes are related to the changed status of the rolling element by crossing over from the diagnostic (load-free) to the loaded part of the functional sample.

The time period of the major and minor amplitudes are approximately equal to the damage time period of the guiding profile: $T_{d42} \cong T_{d31} \cong T_{\text{Dp}}$. Through the time phases, ΔT_{d21} and ΔT_{d43}, between the major and minor amplitudes, the damage position against the current position of the rolling elements can be determined:

$$\Delta T_{d21} = t_{d2} - t_{d1} \tag{24}$$

$$\Delta T_{d43} = t_{d4} - t_{d3} \tag{25}$$

The distance (Δl_d) of the simulated damage against the rolling elements is roughly equal to:

$$\Delta l_d \cong \frac{1}{2}\Delta T_{d21} v \cong \frac{1}{2}\Delta T_{d43} v \tag{26}$$

Through the length of the diagnostic part (l_d), a number of rolling elements (n_d) in contact with one raceway of the guiding profile may be calculated:

$$n_d = \frac{l_d}{d_v} \tag{27}$$

The results of Equations (24)–(27), which are summarised in Table 10, indicate 3 or 4 rolling elements in contact with the raceway at each time (Figure 25).

Table 10. Results of Equations (24)–(27).

Time Phases		Distance of Simulated Damage against Rolling Elements	Length of The Iron Unit	Number of Rolling Elements in Contact with One Raceway of Guiding Profile
$\Delta T_{d21} = 0.0074$ s	$\Delta T_{d43} = 0.0081$ s	$\Delta l_d \cong 1.6$ mm	$l_d = 15$ mm	$n_d = 3.75$

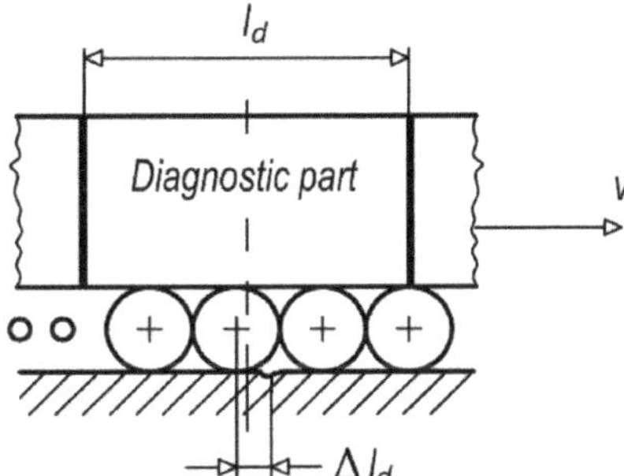

Figure 25. The position of the simulated damage against rolling elements.

Compared with the load-free carriage, the number of rolling elements in contact with the guiding profile is reduced. With consideration to the preload of the linear systems, by crossing one specific rolling element over the damage, the remaining rolling elements might absorb the excited vibrations. Significant changes may be noticed in the time character of the measured accelerations. By reducing the number of the rolling elements in contact with the guiding profile, significant acceleration amplitudes belonging to the vibrations that were excited by the simulated damage were reached.

5. Conclusions

The diagnostics of linear rolling systems is currently based on measuring vibrations and evaluating the RMS value in the context of the threshold value. In transportation practice, we registered several cases of failure without exceeding the threshold value of the vibrations. Therefore, the original principle of diagnostics was proposed on the basis of minimising the external load through the load-free (diagnostic) part that is integrated into the linear system carriage. Through testing, it was proven that the innovative diagnostic principle enables the early detection of failures, even if the linear rolling system is operated under great external loads.

In the first stage, the vibrations that were measured on the loaded and load-free carriage were compared in the time domain. The main conclusions of the first-stage testing are:

- The external load of linear systems negatively affects the early detection of possible failure;
- In the loaded carriage, the vibrations that were excited by the simulated damage were hidden in the measured vibrations;
- In the load-free carriage, minor amplitudes of vibrations were observed in the measured vibrations.

In the second stage, the functional sample with the integrated diagnostic part was designed and tested. The vibrations that were measured on the loaded carriage and the load-free diagnostic part were compared in the time domain. The main conclusions of the second-stage testing are:

- In the loaded carriage, the vibrations that were excited by the simulated damage were hidden in the measured vibrations;
- In the diagnostic part, the major amplitudes related to the simulated damage were registered;
- The major amplitudes at the diagnostic part were reached by reducing the number of rolling elements in contact with the raceway, and by further reducing the external load as the mass inertia of the diagnostic part.

It should also be noted that the vibrations that were excited by the simulated damage appeared once, at a particular time. In the load-free carriage, theoretically, eighteen acceleration amplitudes related to the damage might be noticed in the measured data, and in the diagnostic part, only four. In general, the vibrations that are associated with possible failure dispose of their nonperiodical character and cannot be easily processed in the frequency domain.

The initial results of the original diagnostic principle of linear rolling systems are introduced in the paper. Further research is needed in order to reach sufficient reliability of the diagnostic function and to obtain an adequate life of the redesigned carriage. The research should be focused on a methodology of early failure identification, with respect to the wear progression, the level of lubrication, and the variable operating conditions, both kinematic and dynamical. An optimised design of the functional sample should be tested for its life and load capacity. Further research should also deal with the automatic processing of the measured vibrations, which should aim at the implementation of innovative diagnostics into transportation practice.

Author Contributions: Conceptualisation, R.J. and L.P.; methodology, R.J. and L.P.; software, R.J.; validation, R.J., L.P. and R.G.; formal analysis, R.G.; investigation, R.J.; resources, R.G. and R.J.; data curation, R.J.; writing—original draft preparation, R.J. and R.G.; writing—review and editing, R.J. and R.G.; visualisation, R.J., L.P. and R.G.; supervision, L.P.; project administration, R.G.; funding acquisition, R.G. All authors have read and agreed to the published version of the manuscript.

Funding: This work is a part of the following projects: VEGA 1/0528/20 and KEGA 029TUKE-4/2021.

Institutional Review Board Statement: Not applicable.

Informed Consent Statement: Not applicable.

Data Availability Statement: Not applicable.

Conflicts of Interest: The authors declare no conflict of interest.

References

1. UNCTAD. United Nations Conference on Trade and Development. 2021. Available online: https://unctad.org/webflyer/review-maritime-transport-2021 (accessed on 9 January 2022).
2. COP26. United Nations Climate Change Conference. 2021. Available online: https://ukcop26.org (accessed on 9 January 2022).
3. ESPO. Top 10 Environmental Priorities of EU Ports 2020. EcoPorts Publications. 2020. Available online: https://www.ecoports.com (accessed on 9 January 2022).
4. IMO (International Maritime Organization). Introduction to IMO. Available online: http://www.imo (accessed on 9 January 2022).
5. Marine Environment Protection Committee (MEPC) 77, IMO. 2021. Available online: https://www.imo.org (accessed on 9 January 2022).
6. Piňosová, M.; Andrejiová, M.; Lumnitzer, E. Synergistic effect of risk factors and work environmental quality. *Qual. Access Success* **2018**, *19*, 154–159.
7. Michaelides, M.P.; Herodotou, H.; Lind, M.; Watson, R.T. Port-2-Port Communication Enhancing Short Sea Shipping Performance: The Case Study of Cyprus and the Eastern Mediterranean. *Sustainability* **2019**, *11*, 1912. [CrossRef]
8. Shiri, S.; Huynh, N. Assessment of U.S. chassis supply models on drayage productivity and air emissions. *Trans. Res. Part D* **2018**, *61*, 174–203. [CrossRef]
9. Gharehgozli, A.; Zaerpour, N.; de Koster, R. Container terminal layout design: Transition and future. *Marit. Econ. Logist.* **2020**, *22*, 610–639. [CrossRef]

10. Puškár, M.; Jahnátek, A.; Kuric, I.; Kádárová, J.; Kopas, M.; Šoltésová, M. Complex analysis focused on influence of biodiesel and its mixture on regulated and unregulated emissions of motor vehicles with the aim to protect air quality and environment. Air Quality. *Atmos. Health* **2019**, *12*, 855–864. [CrossRef]
11. Pástor, M.; Živčák, J.; Puškár, M.; Lengvarský, P.; Klačková, I. Application of Advanced Measuring Methods for Identification of Stresses and Deformations of Automotive Structures. *Appl. Sci.* **2020**, *10*, 7510. [CrossRef]
12. Kulka, J.; Mantic, M.; Fedorko, G.; Molnar, V. Failure analysis concerning causes of wear for bridge crane rails and wheels. *Eng. Fail. Anal.* **2020**, *110*, 104441. [CrossRef]
13. Lumnitzer, E.; Andrejiová, M.; Goga Bodnárová, A. Verification of the impact of the used type of excitation noise in determining the acoustic properties of separating constructions. *Measurement* **2016**, *78*, 83–89. [CrossRef]
14. Liptai, P.; Lumnitzer, E.; Moravec, M.; Piňosová, M. Analysis and Classification of Noise Sources of Conveyor Systems by Sound Visualising on the Postal Package Sorting Line. *Adv. Sci. Technol. Res. J.* **2018**, *12*, 172–176. [CrossRef]
15. Piňosová, M.; Andejiová, M.; Liptai, P.; Lumnitzer, E. Objective and subjective evaluation of the risk physical factors near to conveyor system. *Adv. Sci. Technol. Res. J.* **2018**, *12*, 188–196. [CrossRef]
16. THK Co. Ltd. Condition-Detecting Device, Method, and Program, and Information-Recording Medium. Granted Patent EP1598569B1, 23 August 2011.
17. Chommuangpuck, P.; Wanglomklang, T.; Srisertpol, J. Fault detection and diagnosis of linear bearing in auto core adhesion mounting machines based on condition monitoring. *Syst. Sci. Control. Eng.* **2021**, *9*, 290–303. [CrossRef]
18. Kim, M.S.; Yun, Y.P.; Lee, S.; Park, P. Unsupervised anomaly detection of LM guide using variational autoencoder. In Proceedings of the XIth International Symposium on Advanced Topics in Electrical Engineering, Bucharest, Romania, 28–30 March 2019.
19. Kim, M.S.; Yun, Y.P.; Park, P. An explainable convolutional neural network for fault diagnosis in linear motion guide. *IEEE Trans. Ind. Inform.* **2021**, *17*, 4036–4045. [CrossRef]
20. Feng, H.; Chen, R.; Wang, Z. Feature extraction for fault diagnosis based on wavelet packet decomposition: An application on linear rolling guide. *Adv. Mech. Eng.* **2018**, *10*, 1687814018796367. [CrossRef]
21. THK Co. Ltd. Method for Diagnosing Rolling Guide Device Status. Granted Patent JP6747757B2, 26 August 2020.
22. Schaeffler Technologies Ag. & Co. Kg. Method for Lubricating a Linear Guide. Patent Application DE102017113720A1, 27 December 2018.
23. Jirova, R.; Pesik, P. Dynamical load of linear rolling guides. *MM Sci. J.* **2020**, 3943–3949. [CrossRef]
24. Skoda Auto, A.S. Linear Rolling Guide with Integrated Diagnostic Equipment. Granted Patent CZ308232B6, 3 November 2020.
25. Serweta, W.; Okolewski, B.; Blazejczyk-Okolewska, B.; Czolczynski, K.; Kapitaniak, T. Mirror hysteresis and Lyapunov exponents of impact oscillator with symmetrical soft stops. *Int. J. Mech. Sci.* **2015**, *101–102*, 89–98. [CrossRef]
26. Tsuha, N.A.H.; Nonato, F.; Cavalca, K.L. Formulation of a reduced order model for stiffness on elastohydrodynamic line contacts applied to cam-follower mechanism. *Mech. Mach. Theory* **2017**, *113*, 22–39. [CrossRef]
27. Wei, W.; Yimin, Z.; Changyou, L.; Hao, W.; Yanxun, Z. Effects of wear on dynamic characteristics and stability of linear guides. *Mechanica* **2017**, *52*, 2899–2913. [CrossRef]
28. Zhen, N.; Li, Q. Analysis of stress and fatigue life of ball screw with considering the dimension errors ball. *Int. J. Mech. Sci.* **2018**, *137*, 68–76. [CrossRef]
29. Cheng, D.J.; Park, J.H.; Suh, J.S.; Kim, S.J.; Park, C.H. Effects of frictional heat generation on the temperature distribution in roller linear motion rail surface. *J. Mech. Sci. Technol.* **2017**, *31*, 1477–1487. [CrossRef]
30. Cheng, D.J.; Park, J.H.; Kim, S.J. Improved friction model for the roller LM guide considering mechanics analysis. *J. Mech. Sci. Technol.* **2018**, *32*, 2723–2734. [CrossRef]
31. Yunlong, W.; Wenzhong, W.; Shengguang, Z.; Ziqiang, Z. Effects of raceway surface roughness in an angular contact bearing. *Mech. Mach. Theory* **2018**, *121*, 198–212. [CrossRef]
32. Grega, I.; Grega, R.; Homisin, J. Frequency of free vibration in systems with a power-law restoring force. *Bull. Pol. Acad. Sci. Tech. Sci.* **2021**, *69*, e136723. [CrossRef]
33. Wang, W.; Zhang, Y.; Li, C. Dynamic reliability analysis of linear guides in positioning precision. *Mech. Mach. Theory* **2017**, *116*, 451–464. [CrossRef]
34. Krajnak, J.; Homisin, J.; Grega, R.; Kassay, P.; Urbansk, M. The failures of flexible couplings due to self-heating by torsional vibrations—Validation on the heat generation in pneumatic flexible tuner of torsional vibrations. *Eng. Fail. Anal.* **2021**, *119*, 104977. [CrossRef]
35. Bizarre, L.; Nonato, F.; Cavalca, K.L. Formulation of five degrees of freedom ball bearing model accounting for the nonlinear stiffness and damping of elastohydrodynamic point contacts. *Mech. Mach. Theory* **2018**, *124*, 179–196. [CrossRef]
36. Adams, V.; Askenazi, A. *Building Better Products with Finite Element Analysis*; OnWord Press: Santa Fe, NM, USA, 1999.

Journal of
Marine Science and Engineering

MDPI

Article

AIS-Based Scenario Simulation for the Control and Improvement of Ship Emissions in Ports

Sheng-Long Kao [1,2,3], Wu-Hsun Chung [1,2,3,*] and Chao-Wei Chen [1]

[1] Department of Transportation Science, National Taiwan Ocean University, Keelung 202, Taiwan; slkao@email.ntou.edu.tw (S.-L.K.); s8700122000@gmail.com (C.-W.C.)
[2] The Center of Excellence for Ocean Engineering (CEOE), National Taiwan Ocean University, Keelung 202, Taiwan
[3] The Intelligent Maritime Research Center (IMRC), National Taiwan Ocean University, Keelung 202, Taiwan
* Correspondence: wxc218@email.ntou.edu.tw; Tel.: +886-2-24622192

Abstract: Maritime transport is a major mode of transportation. Over 80% of international freight is carried by this mode. A port is a hub of ships and freight in maritime transport. Because of growing environmental concerns, how to effectively monitor, control, and improve ship emissions in a port has become a challenge for port administrations. This study combines automatic identification systems (AIS), ship emission estimation model (SEEM), geographic information system (GIS) mapping, and a scenario simulation technique to create a ship emission scenario simulation model (SESSM) for mapping and assessing current ship emissions alongside various "what-if" improvement options in a port area. A case study of the Port of Keelung in Taiwan is used to illustrate and verify the proposed model. In this case, the distribution and density of ship carbon emissions are mapped, with the ship berthing status being identified as the primary source of ship emissions. Meanwhile, nine "what-if" scenarios based on various combinations of speed policies and shore power supplies are simulated and analyzed. The results show that the proposed scenario simulation model is an effective tool to assess various "what-if" emission improvement options and to identify key factors for emission reduction. The effect of shore power supply on carbon emission reduction is significantly greater than speed policies. If investment costs are an issue, a balanced emission improvement option is suggested by combining a new speed policy and 50% shore power supply.

Keywords: ship emission; scenario simulation; automatic identification system (AIS); carbon emissions; geographic information system (GIS)

Citation: Kao, S.-L.; Chung, W.-H.; Chen, C.-W. AIS-Based Scenario Simulation for the Control and Improvement of Ship Emissions in Ports. *J. Mar. Sci. Eng.* **2022**, *10*, 129. https://doi.org/10.3390/jmse10020129

Academic Editors: María Isabel Lamas Galdo and Michele Viviani

Received: 25 November 2021
Accepted: 16 January 2022
Published: 19 January 2022

Publisher's Note: MDPI stays neutral with regard to jurisdictional claims in published maps and institutional affiliations.

1. Introduction

Maritime transport has been the primary transportation mode adopted in global trade. Currently, it transports more than 80% of the world's freight trade [1]. In maritime transport, ports play a pivotal role, functioning as a hub of ship activities and freight transport across countries. However, this critical function also renders ports a hub of maritime transport pollution. According to the European Sea Ports Organization (ESPO) [2], the top three of the top ten environmental priorities of EU ports in 2020, (1) air quality, (2) climate change, and (3) energy efficiency (see Figure 1), are related to ship emissions at ports. Furthermore, the latest statement from the 2021 United Nations Climate Change Conference (COP 26) indicates that nearly 200 countries agreed to the Glasgow Climate Pact to keep 1.5 °C alive and finalize the outstanding elements of the Paris Agreement [3]. This means that greenhouse gas (GHG) emissions mitigation, adaptation, and financing will come into force in the near future. Additionally, the International Maritime Organization (IMO) also agreed with COP 26 to accelerate its efforts to reduce GHG emissions. IMO's Marine Environment Protection Committee (MEPC) has begun the revision of the Initial IMO Strategy on reduction in (GHG) emissions from ships [4]. These environmental demands make the monitoring and control of the air pollution in ports a great urgency and a huge challenge not only for ocean carriers but also for port administrations and residents.

Figure 1. Top ten environmental priorities of European ports [2].

The primary sources of air pollution in a port are ocean-going vessels, harbor craft, cargo handling equipment, on-road vehicles, and rail locomotives. Among these sources, vessel emissions are the majority of pollutants [5]. However, pre-existing literature does not propose an effective method for providing instant air emission information of ship activities and "what-if" improvement solutions. This research gap deters the progress of environmental priorities and the control of GHG emissions in the shipping industry. This research aims to develop an instant and effective method able to estimate and to map the emissions of ship activities in a port, as well as to simulate the outcomes of various "what-if" scenarios for decision making if environmental improvement measures are taken.

The fourth IMO GHG study shows that maritime transport emits around 1056 million tons of CO_2 emissions annually and is responsible for about 2.89% of global anthropogenic greenhouse gas GHG emissions [6]. Either for air pollution or GHG control, vessel emissions can simply not be ignored. A considerable amount of literature regarding vessel emissions has been published. For instance, Eyring et al. investigated emissions changes of international maritime shipping from 1950 to 2001. Their results suggest that from 1970 to 2001 the world's merchant fleet increased rapidly. This fact led to a corresponding increase in total fuel consumption and air pollutants [7]. Endresen et al. studied the environmental impacts of increased international maritime shipping and mapped the geographic distribution of global shipping operations. They showed not only the past trends of emissions but also the future impacts of emissions [8]. In contrast with the worldwide perspective, several researchers have focused on ship emissions in a specific region. Leonardi and Browne presented a method for assessing the carbon footprints of maritime transportation. Based on the data analysis from import supply chains involving several countries in Europe, the results discussed logistics and supply chain choices, the influence of trip distance, load factor, and ship speed [9]. Ammar and Seddiek investigated the case of RO-RO cargo vessels operating in the Red Sea. They compared the environmental and economic performance of four emission reduction methods based on different fuel combinations for ship emission control [10]. Dragović et al. estimated and analyzed ship emission inventories and externalities in the associated cruise bays and ports of Dubrovnik (Croatia) and Kotor (Montenegro) along the eastern coast of the Adriatic Sea. The work also examined port policies for the effective control of air pollutions in such environmentally sensitive areas [11].

In the relevant studies of ship emissions, the method of measuring and estimating ship emissions is a crucial issue and has been widely investigated. Agrawal et al. measured the emissions of the main propulsion engine, auxiliary engine, and an auxiliary boiler on a crude oil tanker and presented a set of emission factors of pollutants. This work provided valuable measurement information for successive studies of ship emission estimation [12]. Corbett et al. adopted a profit-maximizing equation to estimate economically-efficient ship speeds and discussed the policy impacts of a fuel tax and a speed reduction mandate on carbon emissions [13].

Instead of these aforementioned works which rely on static historical shipping statistics for estimating ship emissions, Zaman et al. [14] analyzed realtime ship data in different operational statuses and developed an algorithm to identify the optimum ship speed with the least fuel consumption and carbon emissions.

In recent years, researchers have developed another type of emissions estimation method based on the automatic identification system (AIS). AIS is a tracking system that has been widely used on vessels and can generate navigational data. This type of method adopts the data automatically collected from AIS to estimate the emissions from vessel activities [15–18]. The primary advantage of the AIS-based method is that an AIS can provide approximately realtime navigational information, which can be applied to other fields, such as emissions monitoring and emissions mapping. The method does not require collecting massive historical shipping statistics in advance. In addition, AIS has been widely installed on various vessels, making additional equipment investment not required. Li et al. [19] presented a high-resolution ship emission inventory for the Pearl River Delta region and showed low uncertainty in utilizing AIS data to improve ship emissions estimates. Chen et al. [20] presented a comprehensive national-scale ship emission inventory in China for 2014 based on AIS data for the full year of 2014.

As mentioned above, vessel emissions are a major source of air pollution in a port. The AIS-based method has been applied to the research on port emissions. For instance, Ng et al. [21] used AIS data to investigate the marine emissions in the neighborhood of Hong Kong and the Pearl River Delta and discussed a potential policy change based on the revealed results. Tichavska and Tovar [22] adopted AIS data to estimate the exhaust pollutants related to ferry and cruise operations by sea in Las Palmas Port. Chen et al. [23] presented a high temporal-spatial ship emission inventory in Qingdao Port and its adjacent waters, also based on AIS data. In addition, Yang et al. [24] and Toscano et al. [25] performed similar studies but considered local issues for Tianjin port and Naples port, respectively. Zhang et al. [26] also used AIS data to estimate the ship emission inventory but focused on unidentified vessels with missing ship parameters. Furthermore, Huang et al. [27] dealt with the needs of real-time ship emission monitoring. They presented a method of dynamic calculation of ship emissions based on real-time ship trajectory data. Weng et al. [28] provided higher spatial-temporal resolution for ship emission estimation.

To date, most of the prior research about maritime emissions based on AIS data focused on macro-scale spatial distribution of ship emissions around the globe or in a broad area of sea or coast, such as the Pearl River Delta, the Yangtze River, the Baltic Sea area, or the Adriatic Sea. Few studies have looked at micro-scale spatial distribution ship emissions in a port. Prior research tends to analyze the existing static condition of air pollution from vessel activities and lacks useful tools to evaluate the emissions in various "what-if" scenarios to identify an appropriate improvement alternative. Instead of analyzing and assessing port emissions in a passive manner, this study introduces the technique of simulation to explore proactive emission improvement alternatives. Few past studies address this issue.

This paper combines historical AIS data, a ship emissions estimation model, and a geographic information system (GIS) to create a scenario simulation model for mapping and assessing ship emissions in a port area. The proposed model can present the distribution and volume of ship emissions not only at current status but also in various "what-if" scenarios. This provides the advantage of realtime environmental monitoring and allows port administration to evaluate which emission improvement alternative performs better.

The rest of this paper is organized as follows: Section 2 describes the study's framework and method, and Section 3 details the case for analysis and simulation. Section 4 discusses the simulation results using the proposed method. Finally, Section 5 presents the conclusions and potential opportunities for future research.

2. Methods

The framework of the proposed method is called Ship Emissions Scenario Simulation Model (SESSM) and is illustrated in Figure 2. It requires three types of input data: ship specifications, AIS data, and port mapping information. Ship specifications include ship size, ship tonnage, and propulsion machine. AIS dynamic data mainly include ship direction, position, speed, etc. These two types of data are the input of the Ship Emissions Estimation Model (SEEM). SEEM uses the data as the basic parameters to estimate the volume of ship emissions. Port mapping information provides the scope of the mapping area of the port for ship emissions monitoring. It is the input of GIS mapping of ship emissions and thus needs to be defined clearly. Combining the output of SEEM and port mapping information, GIS maps the distribution and density of ship emissions in a specific port area. These components form the basic framework of the SESSM. The output of the SESSM can be used either for the illustration of the current ship emission status or for the simulation of different "what-if" scenarios. Furthermore, they can be compared and analyzed to improve the control of ship emissions in a port area and to find appropriate improvement options for ship emission reduction. More details of the framework are described below.

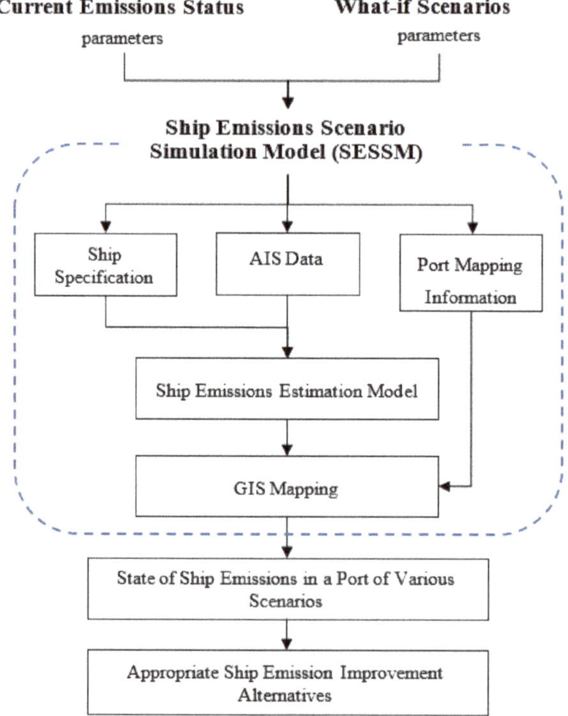

Figure 2. Simulation and evaluation framework.

2.1. Automatic Identification System (AIS)

AIS is a tracking system, which has been widely used on ships at sea. The AIS combines Global Positioning System (GPS) and Very High Frequency (VHF) radio communication technology and enables ships to exchange various navigational information in two different modes—ship-to-ship and ship-to-shore—as shown in Figure 3. The main AIS facilities on land include vessel traffic service (VTS) centers and AIS base stations. The broadcast navigational information of the AIS mainly includes three types: static, dynamic, and voyage. The static information contains ship identification number (known as IMO number), length, beam, and ship type. The dynamic information varies with time, frequently containing position, course, speed, heading, etc. The voyage-related information includes hazardous cargo onboard, draft, destination, route plan, etc. Using this information, the AIS can provide various maritime functions, such as collision avoidance, navigation, maritime security, search, and rescue, etc. In this study, the traditional roles of the AIS are expanded to the environmental monitoring of ship activities.

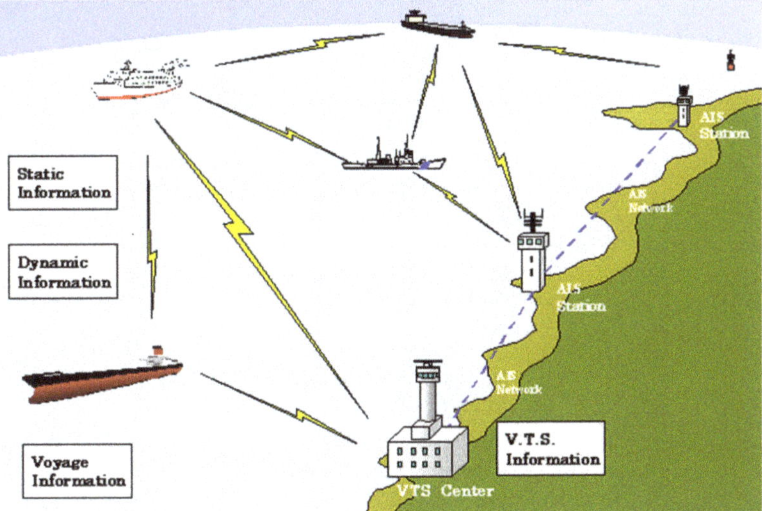

Figure 3. Information exchange in AIS operations [29].

2.2. Ship Emission Estimation Model (SEEM)

Several AIS-based models have been developed to estimate ship emissions [15–18]. Most of these models calculate ship emissions mainly based on engine activities and energy consumption. This study assesses the emissions of individual ships as a function of vessel energy demand multiplied by an emission factor and fuel correction factor as calculated in Equation (1). This estimation model has been implemented and verified by the port of Los Angeles and the major ports of Taiwan [5,30]. The energy demand is the energy output of engines on a ship, which is measured in kW-hr. It comes from three types of sources: main engines, auxiliary engines, and auxiliary boilers. See Equation (2) below. The energy demand is mainly determined by the maximum continuous rated engine power (MCR), load factor (LF), and activity (Act), as shown in Equations (3)–(5). MCR power is defined as the manufacturer's tested engine power and related to the highest power available from a ship engine during average cargo and sea conditions. The load factor means propulsion engine load factor and is expressed as the cube of the ratio of a ship's actual speed to the ship's maximum speed as calculated in Equation (6). From a practical perspective, operating a ship at 100% of its MCR power is very costly in terms of fuel consumption and engine maintenance. Therefore, at normal service speed, a ship usually has a load factor of close to 80%. The activity refers to propulsion engine activity and is measured in operation hours of an engine as calculated in Equation (7). The calculation of the fuel correction factor in Equation (1) follows Table A7 in Appendix A.

$$E = Energy \times EF \times FCF / 10^6 \tag{1}$$

$$Energy = Energy_{me} + Energy_{ae} + Energy_{ab} \tag{2}$$

$$Energy_{me} = MCR \times LF_{me} \times Act \tag{3}$$

$$Energy_{ae} = MCR \times LF_{ae} \times Act \tag{4}$$

$$Energy_{ab} = LF_{ab} \times Act \tag{5}$$

$$LF = \left(\frac{AS}{MS}\right)^3 \tag{6}$$

$$Act = D / AS \tag{7}$$

The nomenclature used in Equations (1)–(7) is provided below.

E: Emission (ton);
$Energy$: Total energy demand (kW-hrs);
$Energy_{me}$: Energy demand of a main engine (kW-hrs);
$Energy_{ae}$: Energy demand of an auxiliary engine (kW-hrs);
$Energy_{ab}$: Energy demand of an auxiliary boiler (kW-hrs);
MCR: Maximum continuous rating power (kW);
LF_{me}: Load factor of a main engine;
LF_{ae}: Load factor of an auxiliary engine;
LF_{ab}: Load factor of an auxiliary boiler;
Act: Activity (hrs);
EF: Emission factor (g/kW-hrs);
FCF: Fuel correction factor;
AS: Actual speed (knots);
MS: Maximum speed (knots); and
D: Distance (nautical miles).

Ship emissions contain various types of pollutants as shown in Appendix A Table A4, such as 10-μm micron particulate matter (PM10), 2.5-μm particulate matter (PM2.5), oxides of nitrogen (NO_x), oxides of sulfur (SO_x), carbon monoxide (CO), etc. Because of intensive concerns on the global impact of the greenhouse effect and climate change in recent years, this study focuses on GHGs, and the results present only one type of emission, carbon dioxide equivalent (CO_{2e}). However, using the Tables A4–A6 in Appendix A, the other pollutants can be easily estimated, and other indicators or multilayer mapping can also be easily applied in the proposed model.

The static information of AIS data, such as IMO ship identification number, can help us identify critical ship characteristics, such as ship tonnage and power sources. Moreover, the dynamic information of AIS data can provide other critical parameters, such as position and speed. These parameters enable SEEM to effectively estimate the emissions of a ship during different times.

2.3. Geographic Information System (GIS)

GIS is a system used to create, analyze, manage, and present various geographic data on a map. It has been widely applied in different fields, including traffic navigation, real estate, national defense, natural resources, etc. Based on the data of ship emissions estimated by SEEM, the GIS in the study is used to visualize the distribution of ship emissions in a port area and to simulate "what-if" scenarios of emissions improvement options. The GIS software used in the paper is ArcMap 10, which maps the port area in grids and plots the density of ship emissions in different colors.

3. Case

3.1. The Port of Keelung

The focus point of this study is the Port of Keelung, a major port in northern Taiwan established in 1886, at 25.1346° N, 121.7411° E (see Figure 4). The Port of Keelung handles about 1.53 million TEU containers and 63 million tons of cargo annually [31]. The AIS data in the study were collected by the AIS base stations around the port in June 2015, including 425 individual ships. Ship status includes sailing on the sea, maneuvering, and berthing. The geographical domain designed for monitoring in the port is the square zone within the range of 20 nautical miles (NM) outside the center of the port, as indicated in Figure 4. The site is plotted in a grid with 500-meter intervals, as shown in Figure 5. The berthing area is the circle area at the bottom of Figure 5.

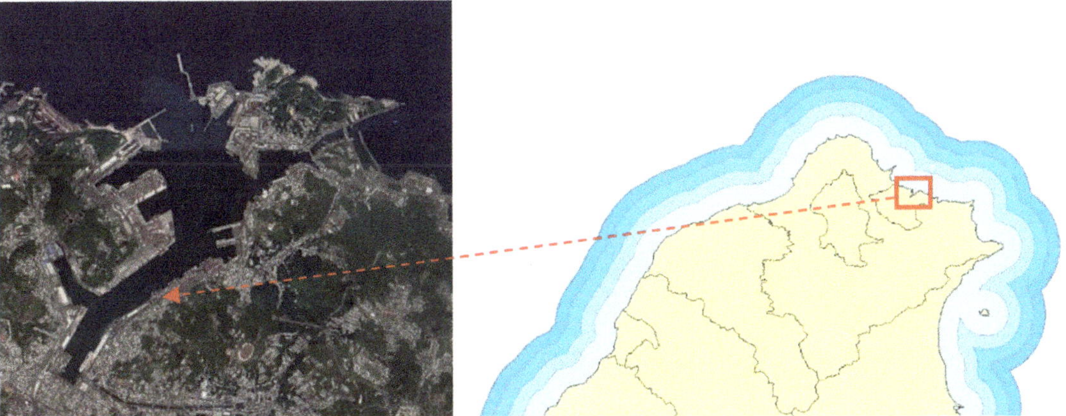

Figure 4. The Port of Keelung in Taiwan (Source: Google Map).

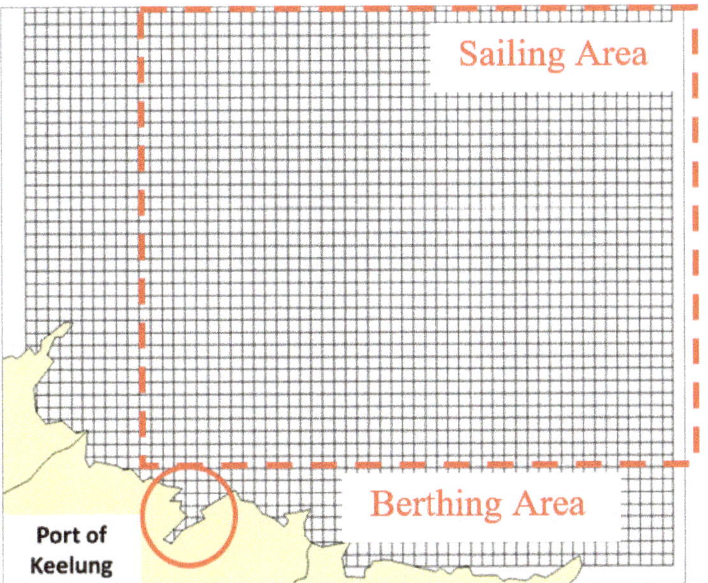

Figure 5. The area of the Port of Keelung designed for monitoring.

3.2. "What-If" Scenarios of Ship Emissions Improvement Plans

In addition to presenting the current status of ship emissions, evaluating various ship emissions improvement plans is another crucial issue in this study. Presently, there are several common measures for vessel air pollution prevention in international port administrations, such as the use of shore power, the use of low-sulfur fuels, the decrease in vessel speed, etc. This study selected vessel speed policy and the use of shore power as key improvement factors with which to construct nine different "what-if" scenarios. Each factor had three options or levels, as shown in Tables 1 and 2.

Table 1. Description of improvement factors.

Improvement Factors	Options/Levels	Remark
Vessel Speed Policy	<20 NM: 12 knots	Current speed policy
	<20 NM: 10 knots <10 NM: 7 knots	Speed policy 1
	<20 NM: 12 knots <15 NM: 10 knots <10 NM: 8 knots <5 NM: 5 knots	Speed policy 2
Shore Power Supply	0% 50% 100%	Current facility status

Table 2. What-if scenarios based on improvement factors.

Scenarios	Improvement Factors	
	Speed Policy	**Shore Power Supply**
1	Current Speed Policy	0%
2	Current Speed Policy	50%
3	Current Speed Policy	100%
4	Speed Policy 1	0%
5	Speed Policy 1	50%
6	Speed Policy 1	100%
7	Speed Policy 2	0%
8	Speed Policy 2	50%
9	Speed Policy 2	100%

In Table 1, the first factor, "Vessel Speed Policy," included three different options: (1) current port speed policy, (2) speed policy 1, and (3) speed policy 2. The "current speed policy" is the speed policy that is being implemented by the Port of Keelung. As shown in Figure 6, it requests the ships to decrease their speed to under 12 knots within the port area (<20 NM). The other two options are "what-if" speed policies for improving emissions. The "speed policy 1" and "speed policy 2" are stepwise speed policies that request ships to decrease the speed at different levels within different distance ranges from the port, as shown in Table 1 and Figures 7 and 8.

In Table 1, the second factor, "Shore Power Supply," has three different levels: 0%, 50%, and 100%. These levels indicate the percentage of onshore power the berthed ships in a port use. Currently, the Port of Keelung has very few facilities providing shore power to berthing ships. Thus, 0% is close to the current status of power supply. The other two levels, 50% and 100%, are "what-if: plans for improving emissions. Because we did not have enough information about how many percentages of berthing ships would turn on their auxiliary boilers, we assumed that berthing ships do not produce emissions for easy estimation. Combining the two factors and their three options (or levels), Table 2 constructs nine "what-if" scenarios. The proposed SESSM can provide the emissions outcomes in the current situation and the "what-if" scenarios.

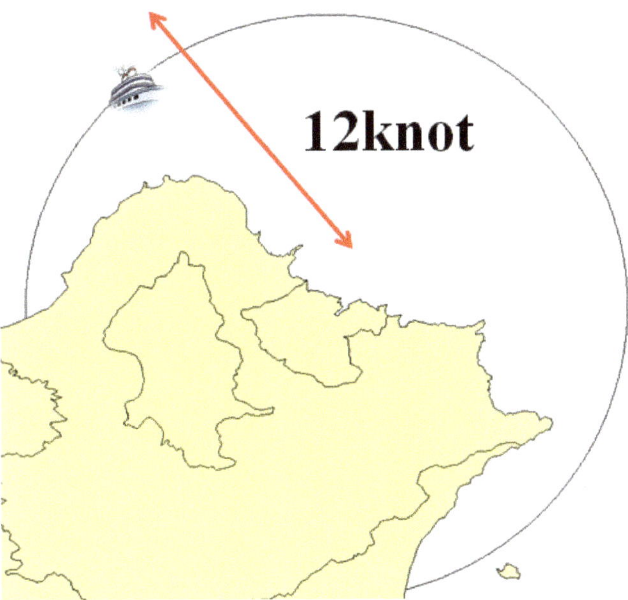

Figure 6. Current speed policy.

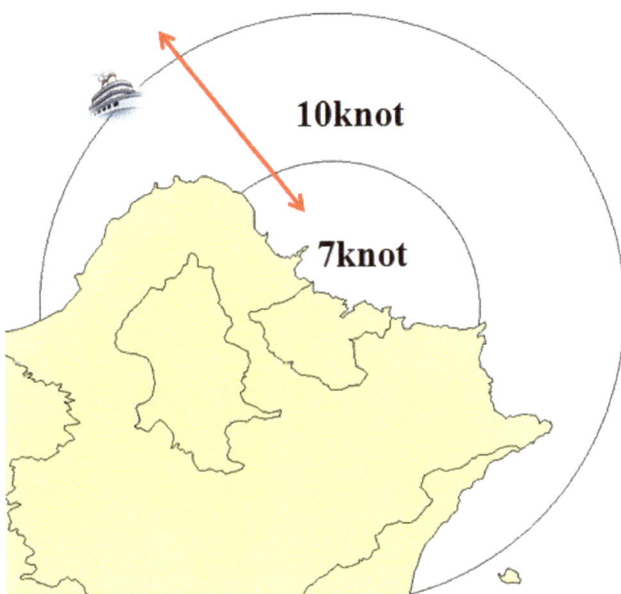

Figure 7. Speed policy 1 (distance interval: 10 NM).

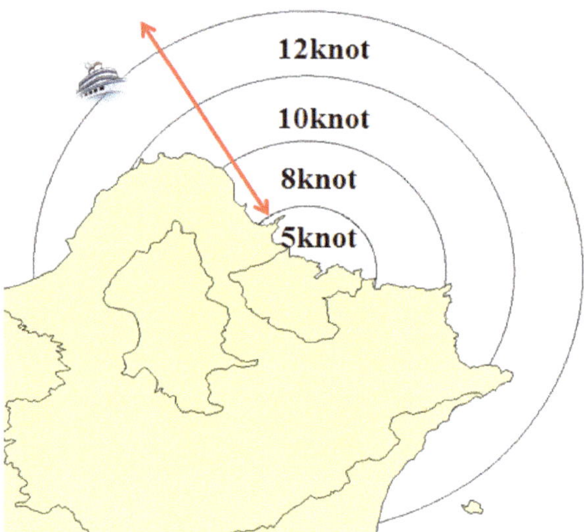

Figure 8. Speed policy 2 (distance interval: 5 NM).

4. Results and Discussions

Based on the proposed SESSM methodology, Figures 9–11 illustrate the simulation result of the distribution and density of carbon emissions in the nine different scenarios in the port area of Keelung. The colors in the grid (500 × 500 m^2) represent the density of carbon emissions. Density is indicated (from low to high) as white, dark green, light green, yellow, orange, and red. The density indicator of carbon emissions (ton per cell in the grid) for different colors is presented in Figure 9.

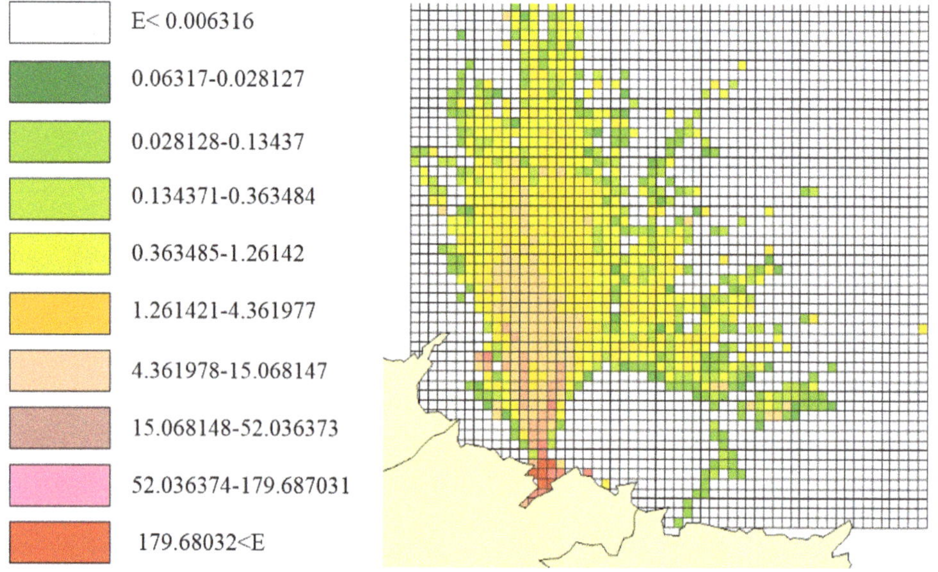

Figure 9. Distribution of carbon emissions in Scenario 1 (current port status).

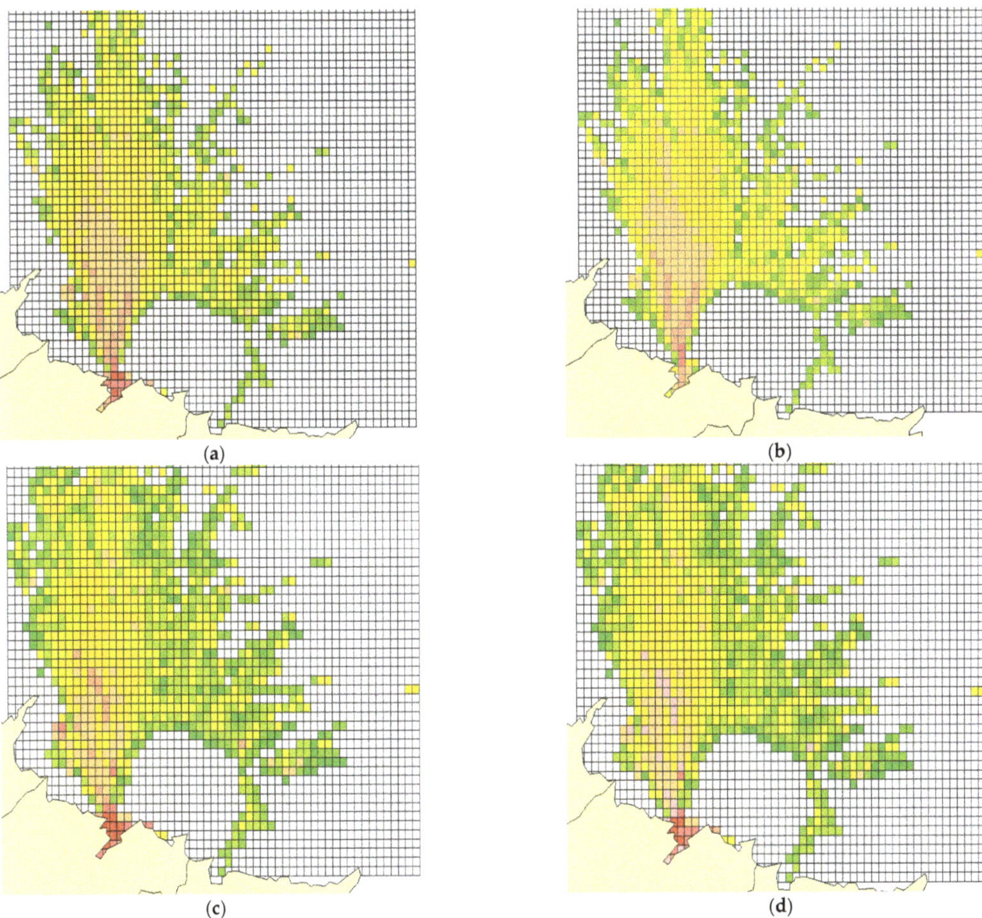

Figure 10. Distribution of carbon emissions in (**a**) Scenario 2 (current speed policy, 50%), (**b**) Scenario 3 (current speed policy, 100%), (**c**) Scenario 4 (speed policy 1, 0%), and (**d**) Scenario 5 (speed policy 1, 50%).

4.1. Vessel Speed Policy

Figures 9 and 10a–d and 11a–d represent three sets of scenarios for the three speed policies. Scenarios 1–3 followed the current speed policy, Scenarios 4–6 followed speed policy 1, and Scenarios 7–9 followed speed policy 2. We can observe that the speed policies have a significant impact on the distribution and density of ship carbon emissions in the sailing area of the port. The sailing area of the port in Figures 9 and 10a,b has more dense red and orange cells than Figures 10c,d and 11a. Figure 11b–d have more dense green and yellow cells than the other figures (i.e., Figures 9, 10 and 11a). This indicates that the proposed stepwise speed policies are environmental friendly, producing less carbon emission than the current speed policy during sailing status. Speed policy 2, which has more interval speed reduction, performs better than speed policy 1 in carbon emissions.

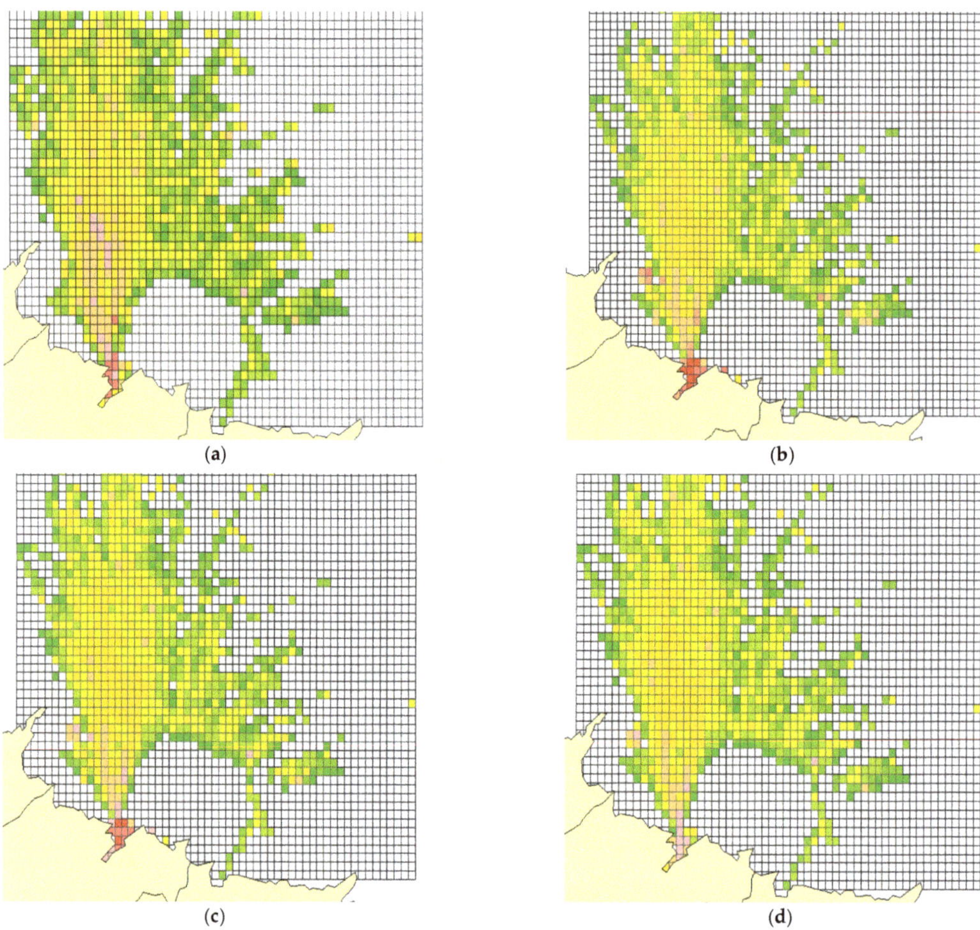

Figure 11. Distribution of carbon emissions in (**a**) Scenario 6 (speed policy 1, 100%), (**b**) Scenario 7 (speed policy 2, 0%), (**c**) Scenario 8 (speed policy 2, 50%), and (**d**) Scenario 9 (speed policy 2, 100%).

4.2. Shore Power Supply

Figures 9, 10c and 11b are the scenarios showing the current port facility status. Figures 10a,d and 11c are the simulation scenarios for shore power levels of 50%, and Figures 10b and 11a,d are the simulation scenarios for shore power levels of 100%. Obviously, the current port status has red cells concentrated in the berthing area, indicating the existing port facilities do not supply any power to berthing ships. It causes ships to produce serious carbon emissions staying in the berthing area. If the shore power level increases to 50%, the number of red cells decreases. This simulation result tells that the ship carbon emissions can be reduced significantly. If the shore power level increases to 100%, the red cells all turn orange or yellow indicating that the ship emissions are improved further. However, the maneuvering activities of the ships in the berthing area still produce a tremendous amount of carbon emissions. This means the cells will not turn completely green.

4.3. Volume of Ship Emissions

Figure 12 shows the total volume of ship carbon emissions and the emissions volume of different ship statuses (sailing and berthing) in various scenarios. Scenario 1, the current port status, has the highest carbon emissions, and Scenario 9 has the lowest carbon emissions. Scenarios 3, 6, and 9 simulate 100% shore power supply, so emissions at the berthing status (grey portion) are zero.

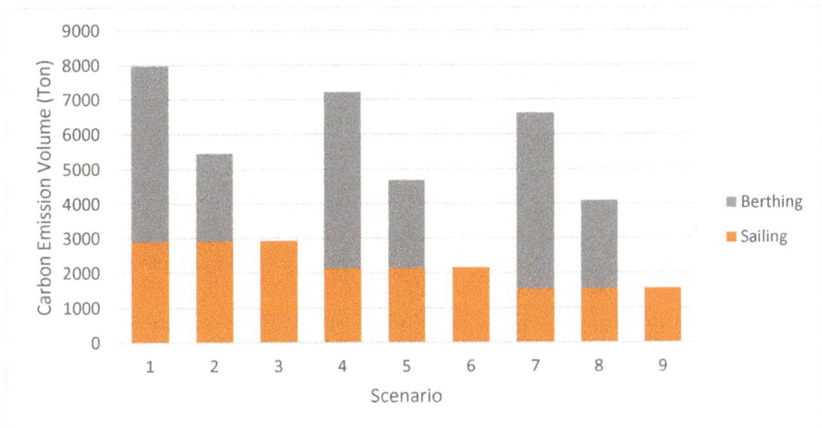

Figure 12. Volume of ship carbon emissions in different scenarios.

Table 3 presents the composition percentage of carbon emissions during sailing and berthing and the influence of the improvement factors, speed policy, and shore power. In the current port status (Scenario 1), carbon emissions generated by sailing and maneuvering are about one-third (36.6%), and those generated by berthing are about two-thirds (63.4%). This indicates that in our mapping area, berthing is the major source of carbon emissions. If the shore power supply increases to 50%, as in Scenario 2, about one-third of the total carbon emissions (31.7%) can be reduced. Suppose the shore power supply increases to 100%, as in Scenario 3, all emissions during berthing can be reduced. That means that two-thirds of the total emission (63.4%) can be reduced simply by using shore power. Therefore, the implementation of shore power is a critical measure for the reduction in carbon emissions in a port area.

Table 3. The influence of improvement factors on ship carbon emissions in different scenarios.

Scenarios	Composition %		Influence Factors		
	Sailing and Maneuvering	Berthing	Speed Policy	Shore Power	Emissions Difference
1	36.6%	63.4%	–	–	current status
2	53.6%	46.4%	–	50%	−31.7%
3	100.0%	0.0%	–	100%	−63.4%
4	29.8%	70.2%	policy 1	–	−9.7%
5	45.9%	54.1%	policy 1	50%	−41.4%
6	100.0%	0.0%	policy 1	100%	−73.1%
7	23.4%	76.6%	policy 2	–	−17.2%
8	37.9%	62.1%	policy 2	50%	−48.9%
9	100.0%	0.0%	policy 2	100%	−80.6%

Because emissions for sailing and maneuvering account for only 36.6%, positive results from a speed policy are much less than those of providing shore power supply. If speed policy 1 is implemented, as outlined in Scenario 4, total emissions can be reduced less than 10%. If speed policy 2 is implemented, as in Scenario 7, less than 18% of total emissions can be reduced. Comparing the two speed policies, the contribution of speed policy 2 to the reduction in total ship carbon emissions is almost double (17.2% vs. 9.7%). Obviously, speed policy 2 outperforms speed policy 1 in emissions reduction.

Presently, the port has insufficient shore power facilities. Since the installation of shore power facilities would require an extra investment cost, increasing the shore power supply to 100% may not be achievable in a short time. Thus, an initial change to 50% shore power supply is a reasonable improvement target. Scenario 8, the improvement option combining speed policy 2 and shore power 50%, is the best one among all 50% shore power options, as it can reduce the most carbon emissions—almost half of the total emissions (48.9%).

5. Conclusions

This paper combined AIS, SEEM, GIS mapping, and a scenario simulation technique to construct a ship emissions scenario simulation model for mapping and assessing the ship emissions of the current status and "what-if" improvement scenarios in a port area. The proposed model successfully mapped and estimated the distribution and density of the Port of Keelung and simulated the other "what-if" improvement scenarios. The results show that SESSM is an effective tool to assess various "what-if" emission improvement options and is able to identify key factors for emission reduction. Based on the case study of the Port of Keelung, the primary source of ship carbon emissions comes from ship berthing status. Thus, the improvement of shore power supply can reduce total ship emissions significantly, especially in the area of the berthing docks. However, this improvement incurs a great number of investment costs. The change of speed policies affects emissions less than the shore power supply does but will not require additional investment costs from port administrations. The improvement option balancing the two factors seems to be the best initial option.

Since the proposed simulation model is innovative to the relevant study of ship emissions control, it may not be sufficiently refined. Many issues have not been fully addressed and need to be perfected in future work. For instance, the simulation model is deterministic. Other critical variables, such as investment costs, operation costs, maintenance costs, weather, and sea conditions have not been considered. A complicated simulation model involving these stochastic and realistic elements can be developed to provide further financial analysis for port planning evaluation. In addition, the scenarios include only two improvement factors—speed policies and shore power supply. If relevant data iare available, more experiment factors and levels can be added into the simulation scenarios to provide port administrations with more feasible and flexible options for decision making.

Author Contributions: S.-L.K. made contributions to the acquisition of data, and the interpretation of results. W.-H.C. conceptualized, designed, drafted and revised the work. C.-W.C. analyzed the data and created graphs and tables. All authors have read and agreed to the published version of the manuscript.

Funding: This research received funding from the Center of Excellence for Ocean Engineering (CEOE) of National Taiwan Ocean University (NTOU), the Intelligent Maritime Research Center (IMRC) of NTOU, and the Ministry of Science and Technology (MOST), Taiwan, R.O.C. under Grant MOST 105-2221-E-019-058 and 110-2221-E-019-071.

Institutional Review Board Statement: Not applicable.

Informed Consent Statement: Not applicable.

Data Availability Statement: Data sharing is not applicable to this article.

Conflicts of Interest: The authors declare no conflict of interest.

Appendix A

Table A1. Parameter Defaults for Ocean-going Vessels [32].

Vessel Type	Maximum Rotational Speed (rpm)	Maximum Main Engine Power (kW)	Maximum Speed (Knot)	Auxiliary Engine Power (kW)
Bulk Carrier	123	8373	14	1486
Container ship	107	32,082	21	6100
Passenger	174	21,848	19	6752
General Cargo	178	4540	13	1195
Ro/Ro	159	8805	19	1175
Tanker	156	7055	14	2179
Other	171	4934	12	1455

Table A2. Auxiliary Engine Power and Load Factor Defaults [32].

Vessel Type	Auxiliary Engine Power (kW)	Load Factor Defaults (%)		
		Sea	Maneuvering	Berthing
Bulk	2850	17%	45%	10%
Container 1000	2090	13%	50%	18%
Container 2000	4925	13%	50%	22%
Container 3000	5931	13%	50%	22%
Container 4000	7121	13%	50%	18%
Container 5000	11,360	13%	50%	16%
Container 6000	13,501	13%	50%	15%
Container 7000	13,501	13%	50%	15%
Container 8000	13,501	13%	50%	15%
Passenger	3900	15%	45%	32%
General Cargo	1776	17%	45%	22%
Ro/Ro	2850	15%	45%	26%
Tanker All_Small	1911	24%	33%	26%
Tanker Panamax	2520	24%	33%	26%
Tanker Afranax	2544	24%	33%	26%
Tanker Suezmax	2865	24%	33%	26%
Tanker VLCC	3388	24%	33%	26%
Tanker ULCC	3667	24%	33%	26%
Other	1776	17%	45%	22%

Table A3. Auxiliary Boiler Load Defaults (kW) [32].

Vessel Type	Auxiliary Boiler Load Defaults (kW)		
	Sea	Maneuvering	Berthing
Bulk	0	109	109
Container	0	506	506
Passenger	0	1000	1000
General Cargo	0	106	106
Ro/Ro	0	109	109
Tanker	0	371	3000
Tanker	0	346	346
Other	0	371	371

Table A4. Main Engine Emission Factors (Unit: g/kWh) [32].

Model Year	NOx	VOC	CO	SO_2	PM10	PM2.5	DPM	CO_2	N_2O	CH_4
<=1999	14	0.5	1.1	11.5	1.5	1.2	1.5	683	0.031	0.01
2000–2010	13	0.5	1.1	11.5	1.5	1.2	1.5	683	0.031	0.01
2011–2015	10.5	0.5	1.1	11.5	1.5	1.2	1.5	683	0.031	0.01

Table A5. Auxiliary Engine Emission Factors (Unit: g/kWh) [32].

Model Year	NOx	VOC	CO	SO$_2$	PM10	PM2.5	DPM	CO$_2$	N$_2$O	CH$_4$
<=1999	14.7	0.5	1.1	12.3	1	0.8	1	683	0.031	0.008
2000–2010	13	0.5	1.1	12.3	1	0.8	1	683	0.031	0.008
2011–2015	10.5	0.5	1.1	12.3	1	0.8	1	683	0.031	0.008

Table A6. Auxiliary Boiler Emission Factors (Unit: g/kWh) [32].

NOx	VOC	CO	SO$_2$	PM10	PM2.5	DPM	CO$_2$	N$_2$O	CH$_4$
2.1	0.1	0.2	16.5	0.8	0.6	0	970	0.08	0.002

Table A7. Fuel Correction Factor [32].

	NOx	VOC	CO	SO$_2$	PM10	PM2.5	DPM	CO$_2$	N$_2$O	CH$_4$
HFO (2.7%S)	1	1	1	1	1	1	1	1	1	1
HFO (1.5%S)	1	1	1	0.555	0.82	0.82	0.82	1	1	1
MGO (0.5%S)	0.94	1	1	0.185	0.25	0.25	0.25	1		1
MDO (1.5%S)	0.94	1	1	0.555	0.47	0.47	0.47	1		1
MGO (0.1%S)	0.94	1	1	0.037	0.17	0.17	0.17	1		1
MGO (0.3%S)	0.94	1	1	0.111	0.21	0.21	0.21	1		1
MGO (0.4%S)	0.94	1	1	0.148	0.23	0.23	0.23	1		1

References

1. IMO (International Maritime Organization). Introduction to IMO. Available online: http://www.imo.org/en/About/Pages/Default.aspx (accessed on 24 November 2021).
2. ESPO. Top 10 Environmental Priorities of EU Ports 2020. EcoPorts Publications. 2020. Available online: https://www.ecoports.com/publications/top-10-environmental-priorities-of-eu-ports-2020 (accessed on 24 November 2021).
3. COP26. United Nations Climate Change Conference. 2021. Available online: https://ukcop26.org/cop26-keeps-1-5c-alive-and-finalises-paris-agreement/ (accessed on 24 November 2021).
4. Marine Environment Protection Committee (MEPC) 77, IMO. 2021. Available online: https://www.imo.org/en/MediaCentre/MeetingSummaries/Pages/MEPC77.aspx (accessed on 12 December 2021).
5. Starcrest Consulting Group, LLC. Port of Los Angeles Port of Los Angeles Inventory of Air Emission—2014. 2015. Available online: https://kentico.portoflosangeles.org/getmedia/8066ecf3-86a1-4fa8-b1c9-f9726b92be67/2014_Air_Emissions_Inventory_Full_Report (accessed on 18 November 2021).
6. Forth Greenhouse Gas Study, IMO. 2020. Available online: https://www.imo.org/en/OurWork/Environment/Pages/Fourth-IMO-Greenhouse-Gas-Study-2020.aspx (accessed on 12 December 2021).
7. Eyring, V.; Köhler, H.; van Aardenne, J.; Lauer, A. Emissions from international shipping: 1. The last 50 years. *J. Geophys. Res. D Atmos.* **2005**, *110*, D17. [CrossRef]
8. Endresen, Ø.; Dalsøren, S.; Eide, M.; Isaksen, I.S.; Sørgård, E. The environmental impacts of increased international maritime shipping–Past trends and future perspectives. In Proceedings of the Global Forum on Transport and Environment in a Globalising World, Guadalajara, Mexico, 10 November 2008.
9. Leonardi, J.; Browne, M. A method for assessing the carbon footprint of maritime freight transport: European case study and results. *Int. J. Logist. Res. Appl.* **2010**, *13*, 349–358. [CrossRef]
10. Ammar, N.R.; Seddiek, I.S. Eco-environmental analysis of ship emission control methods: Case study RO-RO cargo vessel. *Ocean Eng.* **2017**, *137*, 166–173. [CrossRef]
11. Dragović, B.; Tzannatos, E.; Tselentis, V.; Meštrović, R.; Škurić, M. Ship emissions and their externalities in cruise ports. *Transp. Res. D Transp. Environ.* **2018**, *61*, 289–300. [CrossRef]
12. Agrawal, H.; Welch, W.A.; Miller, J.W.; Cocker, D.R. Emission measurements from a crude oil tanker at sea. *Environ. Sci. Technol.* **2008**, *42*, 7098–7103. [CrossRef] [PubMed]
13. Corbett, J.J.; Wang, H.; Winebrake, J.J. The effectiveness and costs of speed reductions on emissions from international shipping. *Transp. Res. D Transp. Environ.* **2009**, *14*, 593–598. [CrossRef]
14. Zaman, I.; Pazouki, K.; Norman, R.; Younessi, S.; Coleman, S. Utilising real-time ship data to save fuel consumption and reduce carbon emission. In *Shipping in Changing Climates Conference*; Newcastle University: Newcastle, UK, 2016.
15. Perez, H.M.; Chang, R.; Billings, R.; Kosub, T.L. Automatic identification systems (AIS) data use in marine vessel emission estimation. In Proceedings of the 18th Annual International Emission Inventory Conference, Baltimore, MD, USA, 14–17 April 2009.
16. Jalkanen, J.-P.; Brink, A.; Kalli, J.; Pettersson, H.; Kukkonen, J.; Stipa, T. A modelling system for the exhaust emissions of marine traffic and its application in the Baltic Sea area. *Atmos. Chem. Phys.* **2009**, *9*, 9209–9223. [CrossRef]

17. Miola, A.; Ciuffo, B. *Regulating Air Emissions from Ships-The State of the Art on Methodologies, Technologies and Policy Options*; Institute for Environment and Sustainability, European Commission, Joint Research Centre: Luxembourg, 2010.
18. Jalkanen, J.-P.; Johansson, L.; Kukkonen, J.; Brink, A.; Kalli, J.; Stipa, T. Extension of an assessment model of ship traffic exhaust emissions for particulate matter and carbon monoxide. *Atmos. Chem. Phys.* **2012**, *12*, 2641–2659. [CrossRef]
19. Li, C.; Yuan, Z.; Ou, J.; Fan, X.; Ye, S.; Xiao, T.; Shi, Y.; Huang, Z.; Ng, S.K.; Zhong, Z. An AIS-based high-resolution ship emission inventory and its uncertainty in Pearl River Delta region, China. *Sci. Total Environ.* **2016**, *573*, 1–10. [CrossRef] [PubMed]
20. Chen, D.; Wang, X.; Li, Y.; Lang, J.; Zhou, Y.; Guo, X.; Zhao, Y. High-spatiotemporal-resolution ship emission inventory of China based on AIS data in 2014. *Sci. Total Environ.* **2017**, *609*, 776–787. [CrossRef] [PubMed]
21. Ng, S.K.; Loh, C.; Lin, C.; Booth, V.; Chan, J.W.; Yip, A.C.; Li, Y.; Lau, A.K. Policy change driven by an AIS-assisted marine emission inventory in Hong Kong and the Pearl River Delta. *Atmos. Environ.* **2013**, *76*, 102–112. [CrossRef]
22. Tichavska, M.; Tovar, B. Port-city Exhaust Emission Model: An approach to Cruise and Ferry operations in Las Palmas Port. In Proceedings of the IAME Conference, Norfolk, VA, USA, 15–18 July 2014.
23. Chen, D.; Wang, X.; Nelson, P.; Li, Y.; Zhao, N.; Zhao, Y.; Lang, J.; Zhou, Y.; Guo, X. Ship emission inventory and its impact on the PM2. 5 air pollution in Qingdao Port, North China. *Atmos. Environ.* **2017**, *166*, 351–361. [CrossRef]
24. Yang, L.; Zhang, Q.; Zhang, Y.; Lv, Z.; Wang, Y.; Wu, L.; Feng, X.; Mao, H. An AIS-based emission inventory and the impact on air quality in Tianjin port based on localized emission factors. *Sci. Total Environ.* **2021**, *783*, 146869. [CrossRef]
25. Toscano, D.; Murena, F.; Quaranta, F.; Mocerino, L. Assessment of the impact of ship emissions on air quality based on a complete annual emission inventory using AIS data for the port of Naples. *Ocean Eng.* **2021**, *232*, 109166. [CrossRef]
26. Zhang, Y.; Fung, J.C.; Chan, J.W.; Lau, A.K. The significance of incorporating unidentified vessels into AIS-based ship emission inventory. *Atmos. Environ.* **2019**, *203*, 102–113. [CrossRef]
27. Huang, L.; Wen, Y.; Zhang, Y.; Zhou, C.; Zhang, F.; Yang, T. Dynamic calculation of ship exhaust emissions based on real-time AIS data. *Transp. Res. Part D Transp. Environ.* **2020**, *80*, 102277. [CrossRef]
28. Weng, J.; Shi, K.; Gan, X.; Li, G.; Huang, Z. Ship emission estimation with high spatial-temporal resolution in the Yangtze River estuary using AIS data. *J. Clean. Prod.* **2020**, *248*, 119297. [CrossRef]
29. Bright Hub Engineering. Automatic Identification System for Tracking Ships. 2010. Available online: https://www. brighthubengineering.com/seafaring/63389-tracking-ships-using-automatic-identification-system/ (accessed on 12 December 2021).
30. Wang, W.-C. Analytical Utilization of AIS Data in Estimating Fuel Consumption and Shipping Emission. Thesis in Chinese. 2014. Available online: https://hdl.handle.net/11296/7xg9kr (accessed on 12 December 2021).
31. Taiwan International Ports Corporation. Introduction of the Port of Keelung. 2020. Available online: https://kl.twport.com.tw/chinese/cp.aspx?n=2883405E97DCC350 (accessed on 12 December 2021).
32. Starcrest Consulting Group, LLC; Puget Sound Maritime Air Forum. Puget Sound Maritime Air Emissions Inventory Executive Summary. 2012. Available online: https://pugetsoundmaritimeairforum.files.wordpress.com/2016/06/2011pseireportupdate_20130523.pdf (accessed on 18 November 2021).

Journal of
Marine Science and Engineering

Article

Possibilities of Ammonia as Both Fuel and NO$_x$ Reductant in Marine Engines: A Numerical Study

Carlos Gervasio Rodríguez [1], María Isabel Lamas [1,*], Juan de Dios Rodríguez [2] and Amr Abbas [3]

[1] Nautical Sciences and Marine Engineering Department, University of A Coruña, Mendizabal s/n, 15403 Ferrol, A Coruña, Spain; c.rodriguez.vidal@udc.es

[2] Industrial Engineering Department, University of A Coruña, Mendizabal s/n, 15403 Ferrol, A Coruña, Spain; de.dios.rodriguez@udc.es

[3] Department of Mechanical Engineering, Mississippi State University, Starkville, MS 39762, USA; aa2642@msstate.edu

* Correspondence: isabel.lamas.galdo@udc.es

Abstract: Nowadays, the environmental impact of shipping constitutes an important challenge. In order to achieve climate neutrality as soon as possible, an important priority consists of progressing on the decarbonization of marine fuels. Free-carbon fuels, used as single fuel or in a dual-fuel mode, are gaining special interest for marine engines. A dual fuel ammonia-diesel operation is proposed in which ammonia is introduced with the intake air. According to this, the present work analyzes the possibilities of ammonia in marine diesel engines. Several ammonia-diesel proportions were analyzed, and it was found that when the proportion of ammonia is increased, important reductions of carbon dioxide, carbon monoxide, and unburnt hydrocarbons are obtained, but at the expense of increments of oxides of nitrogen (NO$_x$), which are only low when too small or too large proportions of ammonia are employed. In order to reduce NO$_x$ too, a second ammonia injection along the expansion stroke is proposed. This measure leads to important NO$_x$ reductions.

Keywords: ammonia; emissions; decarbonization; marine engines

Citation: Rodríguez, C.G.; Lamas, M.I.; Rodríguez, J.d.D.; Abbas, A. Possibilities of Ammonia as Both Fuel and NO$_x$ Reductant in Marine Engines: A Numerical Study. *J. Mar. Sci. Eng.* **2022**, *10*, 43. https://doi.org/10.3390/jmse10010043

Academic Editor: Tie Li

Received: 27 November 2021
Accepted: 24 December 2021
Published: 1 January 2022

Publisher's Note: MDPI stays neutral with regard to jurisdictional claims in published maps and institutional affiliations.

1. Introduction

Marine transport, mainly powered by diesel engines, accounts for more than 90% of the transport of international trade goods [1]. The climate impact of shipping is one of the most important areas of ecology since ships are responsible for 2.2% of carbon dioxide (CO$_2$), 20.8% of nitrogen oxides (NO$_x$), 11.8% of sulfur oxides (SO$_x$), 8.57% of particulate matter 2.5 (PM2.5), and 4.63% of particulate matter 10 (PM10) emissions worldwide in 2019 [2,3]. Several restrictions have been imposed to reduce emissions from ships. The most crucial ones are included in the 73/78 MARPOL convention, (International Convention for the Prevention of Pollution from Ships), by the International Maritime Organization (IMO). This convention, which came into force in 2005 and is revised periodically, regulates several aspects of marine environmental pollution and has recently proposed a decarbonization strategy. One of the objectives of the IMO is the decarbonization of marine diesel engines as soon as possible along this century. Although renewable natural fuels such as biodiesel are gaining importance [4], carbon-free fuels are crucial to achieving the decarbonization of diesel engines. Two promising fuels which fulfill this requirement are hydrogen (H$_2$) and ammonia (NH$_3$). These fuels do not contain carbon nor sulfur and thus their combustion does not generate carbon emissions (CO$_2$, CO, HC, soot), or SO$_x$. Despite the good performance and low emissions of hydrogen, its storage is too complicated to be employed in marine engines [5]. Nevertheless, storage and distribution of ammonia are much easier. Besides, there is available infrastructure for the storage and transport of ammonia which can be used. Ammonia can be easily liquefied and stored at moderate pressures and temperatures, which makes ammonia easy to store on a ship.

On the other hand, hydrogen needs pressures that are too high at ambient temperature or too low, around 20 K [6]. According to this, ammonia storage is considerably cheaper than hydrogen storage. Another advantage of ammonia is that its manipulation under safe conditions is well documented. On the other hand, the main disadvantages are NH_3 slip and emissions of NO_x and N_2O. Ammonia is toxic, and its high slip concentration leads to risks and eutrophication.

Ammonia was first used as fuel in 1822. Sir Goldsworghy Gurney was the first person who used ammonia as fuel and applied it to a locomotive. The application of ammonia as fuel presents two main periods in history: developing an alternative fuel to face any oil crisis and, in the recent period, protecting the environment. The 1940s was an important decade for ammonia as a fuel due to the shortage of conventional fossil fuels during World War II. After World War II, the goal of engineers was to focus on alternative fuels to face any other possible future oil crisis. After this research, no significant research about ammonia as fuel was developed for a long time. However, ammonia has regained interest in recent years due to environmental reasons. It is worth mentioning that, despite it being applied as a fuel many years ago, the research of ammonia as a fuel for internal combustion engines is still in its infancy.

The literature shows that ammonia can be used in both spark ignition (SI) and compression ignition (CI) engines [7–11]. In SI engines, an important advantage is the high-octane number of ammonia, which improves the combustion properties and knock (when the fuel is abnormally auto ignited in local hot spots). On the other hand, the combustion of ammonia in CI engines is much more difficult due to the high autoignition temperature, narrow flammability limits, low flame speed, and high heat of vaporization. An appropriate performance in CI operation was only achieved with high compression ratios required for the autoignition of the fuel [7]. However, the option of partially replacing diesel fuel with ammonia in a dual fuel operation is a realistic option since diesel can be used to start the combustion of the mixture. Several authors obtained satisfactory combustion when ammonia gas is mixed into the intake air [8–14]. Regarding the marine field, currently, there are no commercial solutions but the main marine engines manufacturers such as MAN B&W, Wärtsilä, Caterpillar, etc., are developing encouraging studies to employ ammonia as fuel [15].

In the present work, ammonia is proposed to be used as a fuel for marine diesel engines. A dual fuel mode was analyzed using CFD (computational fluid dynamics), in which ammonia gas is introduced into the air-intake manifold, while diesel fuel was injected directly into the cylinder to trigger the mixture. Since one of the main drawbacks of ammonia is NO_x production, a second ammonia injection along the expansion stroke is proposed. This second injection leads to important NO_x reductions.

2. Materials and Methods

The engine analyzed in the present work, the MAN D2840LE V10, is a four-stroke diesel engine with 10 V-form cylinders and 18270 cm^3 cylinder displacement volume. Each cylinder has one inlet and one exhaust valve, and the fuel injector is placed at the center of the cylinder head. The main characteristics at 100% load are summarized in Table 1.

Table 1. Engine characteristics.

Parameter	Value
Power (kW)	320
Speed (rpm)	1500
Compression ratio	13.5:1
Injection pressure (bar)	220

Regarding the CFD model, Figure 1 shows the computational mesh at the bottom dead center position. A deforming mesh was employed in order to implement the movement of the valves and pistons.

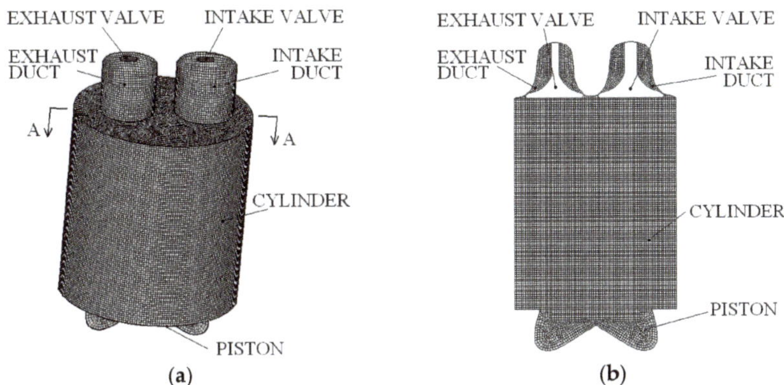

Figure 1. Computational mesh at bottom dead center. (**a**) Tri-dimensional view; (**b**) A-A section.

The CFD simulations were carried out through the open software OpenFOAM. As diesel combustion, ammonia combustion, and NO_x reduction schemes, the models of Ra and Reitz [16], Mathieu and Peterson [17], and Miller and Glarborg [18] were employed, respectively. The fuel heat-up and evaporation were treated through the Dukowicz [19] model and the fuel droplet breakup through the Kelvin-Helmoltz and Rayleigh-Taylor [20] model. The CFD model is based on the equations of conservation of mass, momentum, and energy, Equations (1)–(3), respectively.

$$\frac{\partial \rho}{\partial t} + \frac{\partial}{\partial x_i}(\rho u_i) = 0 \tag{1}$$

$$\frac{\partial}{\partial t}(\rho u_i) + \frac{\partial}{\partial x_j}(\rho u_i u_j) = -\frac{\partial p}{\partial x_i} + \frac{\partial \tau_{ij}}{\partial x_j} + \frac{\partial}{\partial x_j}(-\rho \overline{u_i' u_j'}) \tag{2}$$

$$\frac{\partial}{\partial t}(\rho H) + \frac{\partial}{\partial x_i}(\rho u_i H) = \frac{\partial}{\partial x_i}\left(\frac{\mu_t}{\sigma_h}\frac{\partial H}{\partial x_i}\right) + S_{rad} \tag{3}$$

where ρ represents the density, τ_{ij} the viscous stress tensor, σ_h the turbulent Prandtl number, μ_t the turbulent viscosity, and H the total enthalpy. The chemical reactions were treated through additional equations. Given a set of N species and m reactions, Equation (4), the local mass fraction of each species, f_k, can be expressed by Equation (5).

$$\sum_{k=1}^{N} v'_{kj} M_k \overset{k_j}{\longleftrightarrow} \sum_{k=1}^{N} v''_{kj} M_k \qquad j = 1, 2, \ldots, m \tag{4}$$

$$\frac{\partial}{\partial t}(\rho f_k) + \frac{\partial}{\partial x_i}(\rho u_i f_k) = \frac{\partial}{\partial x_i}\left(\frac{\mu_t}{Sc_t}\frac{\partial f_k}{\partial x_i}\right) + S_k \tag{5}$$

where v'_{kj} are the stoichiometric coefficients of the reactant species M_k in the reaction j, v''_{kj} the stoichiometric coefficients of the product species M_k in the reaction j, Sc_t the turbulent Smidth number and S_k the net rate of production of the species M_k by chemical reaction, given by the molecular weight multiplied by the production rate of the species, Equation (6).

$$S_k = MW_k \frac{d[M_k]}{dt} \tag{6}$$

where MW_k is the molecular weight of the species M_k and $[M_k]$ its concentration. The net progress rate is given by the production of the species M_k minus the destruction of the species M_k along the m reactions:

$$\frac{d[M_k]}{dt} = \sum_{j=1}^{m}\left\{ (v''_{kj} - v'_{kj})\left[k_{fj}\prod_{k=1}^{N}[M_k]^{v'_{kj}} - k_{bj}\prod_{k=1}^{N}[M_k]^{v''_{kj}} \right]\right\} \tag{7}$$

where k_{fj} and k_{fb} are the forward and backward reaction rate constants for each reaction j.

Operating under diesel, the validation with experimental results was developed in previous works [21–23] and thus is not shown here in detail. Figure 2 illustrates the experimentally and numerically obtained in-cylinder pressure against the crank angle at 100% load and Figure 3 the experimentally and numerically obtained SFC (specific fuel consumption) and emissions at several loads. In the experimental tests, the gas analyzers Gasboard-3000 and Gasboard-3030 were employed. Operating under the dual fuel mode ammonia-diesel, the validation using experimental results was not realized due to safety reasons. Ammonia is highly toxic and any accident during the experimental sets could have dramatic consequences for the staff. Nevertheless, the ammonia combustion mechanism was validated for several equivalence ratios, temperatures, and pressures elsewhere [17], in which satisfactory results have been obtained for the species concentrations. NO_x is mainly produced by thermal, fuel and prompt mechanisms, and Mathieu and Peterson found that their model is able to accurately predict NO_x.

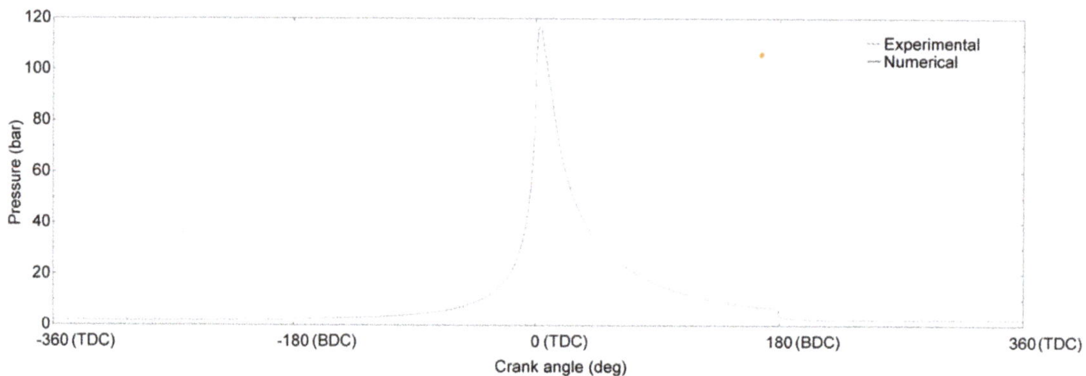

Figure 2. In-cylinder pressure.

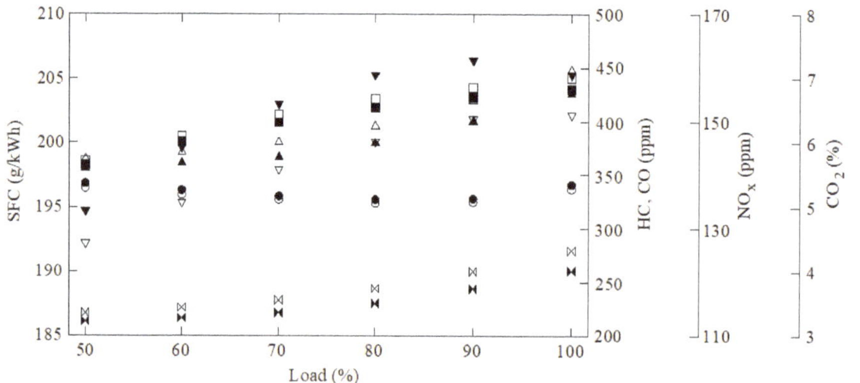

Figure 3. Emissions and consumption.

Regarding Figure 2, the relation between the crank angle and volume is given by Equation (8) [24].

$$V = V_c + \frac{\pi B^2}{4}\left(L + A - A\cos\theta - \sqrt{L^2 - A^2\sin^2\theta}\right) \tag{8}$$

where V is the cylinder volume, V_c is the clearance volume, B the bore, L the connecting rod length, A the crank radius and θ the crank angle.

3. Results and Discussion

Once validated, this CFD model was employed to analyze the dual-mode ammonia-diesel operation. Different proportions of diesel and ammonia were employed under the same power, 320 kW, corresponding to the nominal power using diesel fuel alone. The intake energy rate from diesel or ammonia fuel was computed by the fuel flow rate multiplied by the lower heating value. According to this, the energy contribution from diesel fuel can be obtained by the following expression:

$$\dot{E}_{diesel} = \frac{\dot{m}_{diesel}LHV_{diesel}}{\dot{m}_{diesel}LHV_{diesel} + \dot{m}_{NH_3}LHV_{NH_3}} \tag{9}$$

where $\dot{m}_{diesel}$ and $\dot{m}_{NH_3}$ are the fuel flow rate of diesel and ammonia, respectively, while LHV_{diesel} and LHV_{NH3} are the lower heating value of diesel and ammonia, respectively.

Several experimental results available in the literature concluded that a 100% ammonia fuel mode (i.e., $\dot{E}_{diesel} = 0$) leads to performance problems in compression ignition engines because ammonia has a high resistance to autoignition [25,26]. According to this, the power contribution from diesel fuel analyzed in the present work encompasses the range from 10 to 100%. It is worth mentioning that it is possible to reach 320 kW using only 10% diesel power contribution but at expenses of an excessive ammonia fuel contribution and thus considerable emissions of non-reacted ammonia to the exhaust gas. Nevertheless, these low proportions of power contribution from diesel fuel were also analyzed for illustrative purposes. Regarding NO_x, it is well-known that almost all NO_x produced by compression ignition engines is NO [22,27]. The NO emissions obtained in the present work against the power contribution from diesel fuel are shown in Figure 4. On the one hand, ammonia promotes NO emissions due to its nitrogen content. On the other hand, ammonia leads to lower combustion temperatures (Figure 5 illustrates the in-cylinder average temperature under 50% and 100% power contribution from diesel fuel). Since the main source of NO in internal combustion engines are the high temperatures that are reached in the cylinder [28–31], a reduction of these temperatures leads to a NO emission reduction too. These opposed effects between promoting/mitigating NO formation by ammonia fuel make the net result unpredictable and are responsible for the pattern shown in Figure 4. Under low power contributions from diesel fuel, the quantity of ammonia introduced into the cylinder is so excessive that the NO emissions are high. As the power contribution from diesel fuel is increased, the quantity of ammonia is reduced, and thus NO emissions. At around 70% power contribution from diesel fuel, the NO emissions are minimal and these slightly increase again when the power contribution from diesel fuel is increased due to the increment of the combustion temperature.

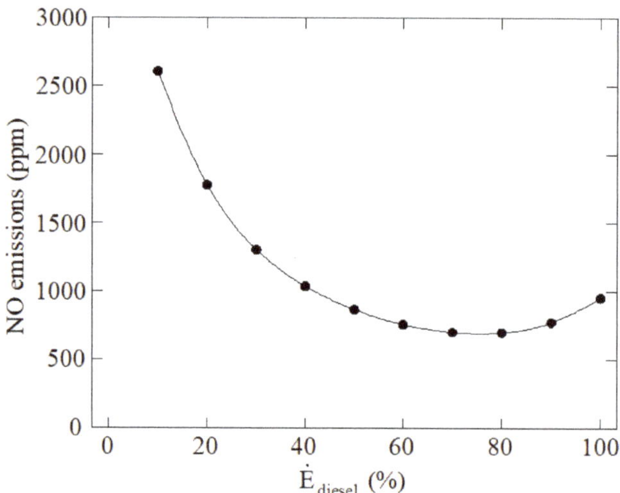

Figure 4. NO emissions against the power contribution from diesel fuel.

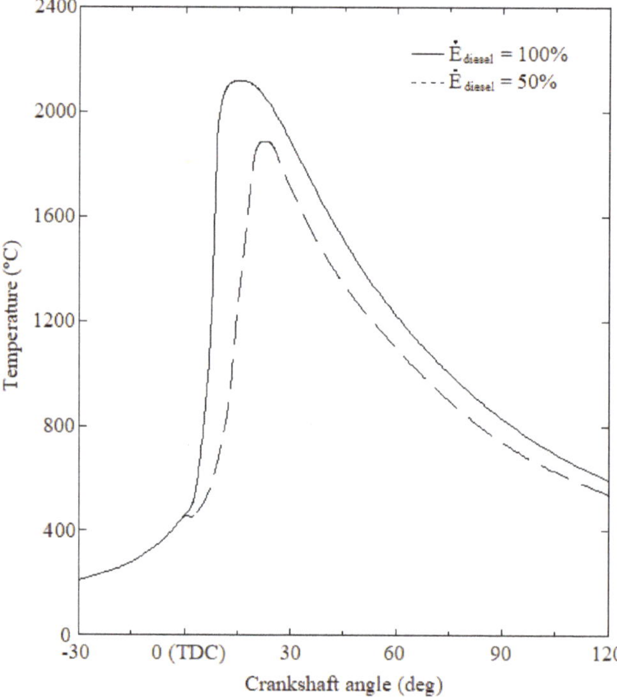

Figure 5. In-cylinder average temperature under 100% and 50% power contribution from diesel fuel.

Regarding CO and HC emissions, these are shown in Figures 6 and 7, respectively. As in the case of NO, two opposed effects can also be found. On the one hand, high quantities of ammonia promote less CO and HC emissions due to the lower carbon content of the whole fuel, since ammonia has no carbon. On the other hand, the lower combustion temperatures obtained when using ammonia promote incomplete combustion and thus lead to CO and HC formation. HC and CO are produced mainly by the slow combustion and partial burning caused by lower combustion temperatures. These opposing effects can be shown in Figures 6 and 7. As can be seen, when the power contribution from diesel fuel is low the CO and HC emissions are low too, these increment to a maximum value corresponding to around 50% power contribution from diesel fuel. From this power contribution, the CO and HC emissions decrease again.

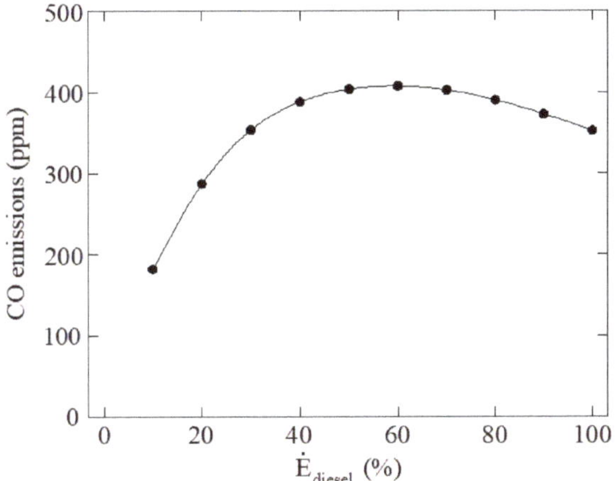

Figure 6. CO emissions against the power contribution from diesel fuel.

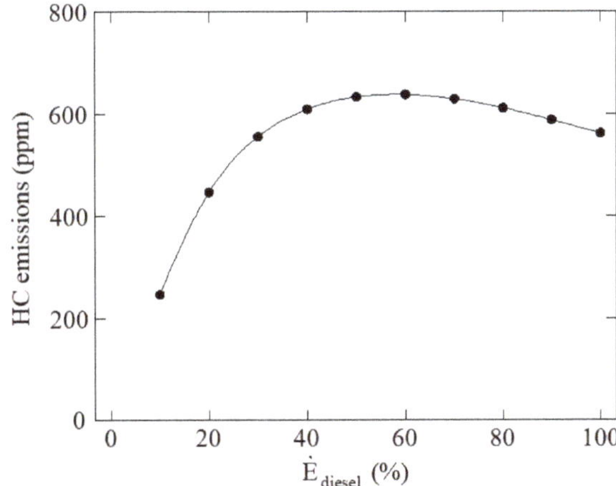

Figure 7. HC emissions against the power contribution from diesel fuel.

CO_2 emissions are illustrated in Figure 8. These emissions are clearly reduced as diesel fuel contribution is reduced too due to the lower carbon content when more ammonia and less diesel is employed as fuel. The relation between the power contribution from diesel fuel and CO_2 emissions is not linear because when the power contribution from diesel fuel is low the proportion of ammonia in the fuel is too high to reach the 320 kW established. The heating value of ammonia, 18.6 MJ/kg, is considerably lower than the heating value of diesel, 42.4 MJ/kg. According to this, CO_2 is drastically reduced under low power contributions from diesel fuel since the quantity of diesel fuel is much lower than the quantity of ammonia fuel.

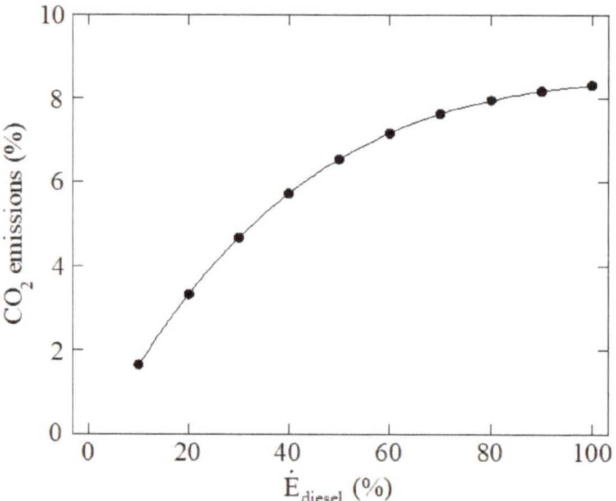

Figure 8. CO_2 emissions against the power contribution from diesel fuel.

The thermodynamic efficiency is shown in Figure 9. As can be seen, this efficiency decreases as more ammonia is employed. The main reason is that ammonia does not burn completely, and a proportion of the intake ammonia is emitted with the exhaust gases. Another reason is that as more ammonia is employed the compression ratio is reduced. Since the efficiency is related to the compression ratio, this is another reason that explains the reduction of thermodynamic efficiency when using ammonia. The ammonia utilization efficiency is also represented in Figure 9, which was computed through the following expression:

$$\eta_{NH_3} = 1 - \frac{\dot{m}_{NH_3 output}}{\dot{m}_{NH_3 input}} \tag{10}$$

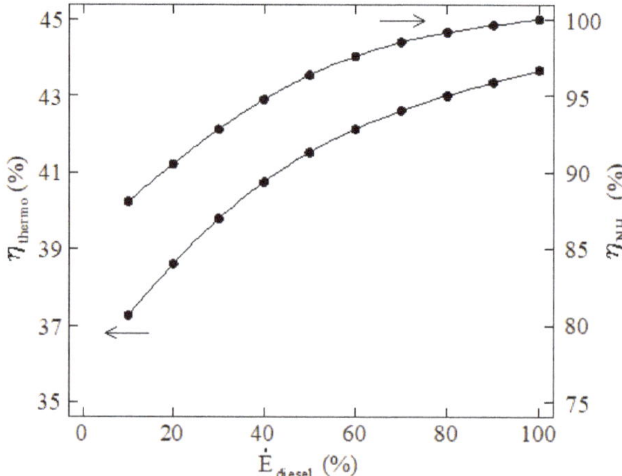

Figure 9. Thermodynamic and ammonia utilization efficiencies against the power contribution from diesel fuel.

Figure 10 shows the in-cylinder pressure against the whole cycle for 50% and 100% diesel contribution. As can be seen, the pressure is reduced when ammonia is employed.

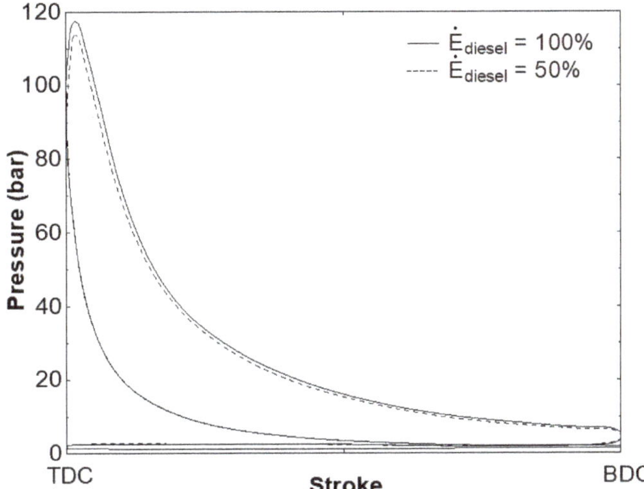

Figure 10. In-cylinder pressure under 100% and 50% power contribution from diesel fuel.

Besides fuel, another application of ammonia consists of NO_x reduction agent. Ammonia is highly employed in SCR (Selective Catalytic Reduction) and SNCR (selective non-catalytic reduction) post-treatments. The main disadvantage of SNCR is that this procedure is only efficient in a narrow temperature range, around 1100–1400 K, considerably higher than the usual temperatures of flue gas from diesel engines. In SCR, the NO_x reduction can be realized at the common temperatures of flue gas from diesel engines by the use of catalysts. The main disadvantages of SCR are the price and limited durability of catalysts. According to this, the present work proposes to realize an additional ammonia injection along the expansion stroke, when the in-cylinder temperature is optimal for NO_x reduction. For instance, the low NO emissions obtained for 70% power contribution from diesel fuel in Figure 4 can be further reduced by injecting ammonia along the expansion stroke. Particularly, in Figure 8 ammonia was injected at 40° crankshaft angle after top dead center, and several ammonia to initial NO (NO_i) ratios were analyzed. As can be seen in Figure 9, NO emissions are highly reduced, but it is worth mentioning the increment of non-reacting ammonia into the exhaust gas when too much ammonia is employed. This non-reacting ammonia is called ammonia slip. According to this, this measure requires precise control of the ammonia injection rate to avoid too high ammonia emissions to the atmosphere. This problem is also characteristic of SCR systems, which also require precise control of the ammonia injected. Another handicap of ammonia consists of the formation of N_2O, which has a high global warming effect. Figure 11 also includes N_2O emissions, which increase as more ammonia is introduced into the engine.

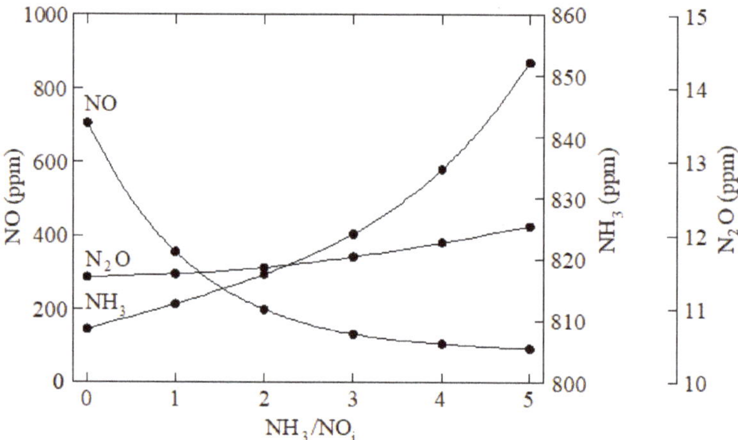

Figure 11. NO, N$_2$O, and NH$_3$ emissions against the NH$_3$/NO$_i$ ratio.

4. Conclusions

The present work focuses on ammonia application as an alternative fuel to be used in compression ignition engines, especially marine engines. The goal is the decarbonization of the marine field. A dual fuel mode based on ammonia injection with the intake air was modeled through CFD. Different proportions of diesel and ammonia were analyzed. It was found that the thermodynamic efficiency was reduced as more ammonia is employed, which leads to a higher engine energy consumption when using ammonia compared to diesel fuel. As expected, a considerable reduction of CO$_2$ was obtained when using ammonia since this fuel does not contain carbon. Regarding CO and HC emissions, too high ammonia proportions led to low CO and HC emissions. On the other hand, too low ammonia proportions also reduce CO and HC emissions due to the higher combustion temperatures, which facilitates complete combustion. Regarding NO$_x$ emissions, two opposed effects were also found. On the one hand, low proportions of ammonia lead to high NO$_x$ emissions due to the high combustion temperatures. On the other hand, high proportions of ammonia lead to high nitrogen introduced into the engine and thus high NO emissions. These two opposing effects caused decreasing/increasing NO emissions depending on the ammonia proportion. Since high ammonia proportions lead to extremely high NO emissions, a measure to reduce these emissions was proposed in the present work. This measure consists of injecting ammonia again during the expansion stroke in order to act as NO reductant. Considerable NO reductions were obtained through this measure but at expenses of excessive NH$_3$ slip to the exhaust gas (which exceeds safe limits) and N$_2$O emissions. In order to be a viable alternative, the ammonia injection procedure must be carefully analyzed and/or after-treatment systems since NH$_3$ is highly toxic and N$_2$O is a powerful greenhouse gas, much more dangerous than carbon dioxide. Future works will focus on reducing both NH$_3$ and N$_2$O emissions.

Author Contributions: Conceptualization, C.G.R., M.I.L., J.d.D.R. and A.A.; methodology, C.G.R., M.I.L., J.d.D.R. and A.A.; software, C.G.R., M.I.L., J.d.D.R. and A.A.; validation, C.G.R.; formal analysis, M.I.L.; investigation, C.G.R., M.I.L., J.d.D.R. and A.A.; resources, C.G.R.; data curation, C.G.R.; writing—original draft preparation, M.I.L.; writing—review and editing, C.G.R., J.d.D.R. and A.A.; supervision, M.I.L., J.d.D.R. and A.A.; project administration, C.G.R., M.I.L., J.d.D.R. and A.A. All authors have read and agreed to the published version of the manuscript.

Funding: This research received no external funding.

Informed Consent Statement: Not applicable.

Acknowledgments: The authors would like to express their gratitude to Norplan Engineering S.L. and recommend the courses "CFD with OpenFOAM" and "C ++ applied to OpenFOAM" available at www.technicalcourses.net (accessed on 26 November 2021).

Conflicts of Interest: The authors declare no conflict of interest.

References

1. Deng, J.; Wang, X.; Wei, Z.; Wang, L.; Wang, C.; Chen, Z. A review of NOx and SOx emission reduction technologies for marine diesel engines and the potential evaluation of liquefied natural gas fuelled vessels. *Sci. Total Environ.* **2021**, *766*, 144319. [CrossRef] [PubMed]
2. Zincir, B. A short review of ammonia as an alternative marine fuel for decarbonised maritime transportation. In Proceedings of the ICEESEN2020, Kayseri, Turkey, 19–21 November 2020.
3. European Energy Agency (EEA). Emission of Air Pollutants from Transport. Available online: https://www.eea.europa.eu/data-and-maps/indicators/transport-emissions-of-air-pollutants-8/transport-emissions-of-air-pollutants-6#tab-related-briefings (accessed on 26 November 2021).
4. Puškár, M.; Kopas, M.; Sabadka, D.; Kliment, M.; Šoltésová, M. Reduction of the gaseous emissions in the marine diesel engine using biodiesel mixtures. *J. Mar. Sci. Eng.* **2020**, *8*, 330. [CrossRef]
5. Seddiek, I.S.; Elgohary, M.M.; Ammar, N.R. The hydrogen-fuelled internal combustion engines for marine applications with a case study. *Brodogradnja* **2015**, *66*, 23–38.
6. Dimitriou, P.; Javaid, R. A review of ammonia as a compression ignition engine fuel. *Int. J. Hydrog. Energy.* **2020**, *45*, 7098–7118. [CrossRef]
7. Lee, D.; Song, H.H. Development of combustion strategy for the internal combustion engine fueled by ammonia and its operating characteristics. *J. Mech. Sci. Technol.* **2018**, *32*, 1905–1925. [CrossRef]
8. Reiter, A.J.; Jong, S.C. Demonstration of compression-ignition engine combustion using ammonia in reducing greenhouse gas emissions. *Energy Fuels.* **2008**, *22*, 2963–2971. [CrossRef]
9. Reiter, A.J.; Kong, S.C. Combustion and emissions characteristics of compression-ignition engine using dual ammonia-diesel fuel. *Fuel* **2011**, *90*, 87–97. [CrossRef]
10. Niki, Y.; Nitta, Y.; Sekiguchi, H.; Hirata, K. Diesel fuel multiple injection effects on emission characteristics of diesel engine mixed ammonia gas into intake air. *J. Eng. Gas Turbines Power* **2019**, *141*, 061020. [CrossRef]
11. Gill, S.S.; Chatha, G.S.; Tsolakis, A.; Golunski, S.E.; York, A.P.E. Assessing the effects of partially decarbonising a diesel engine by co-fuelling with dissociated ammonia. *Int. J. Hydrog. Energy* **2012**, *37*, 6074–6083. [CrossRef]
12. Niki, Y.; Nitta, Y.; Sekiguchi, H.; Hirata, K. Emission and combustion characteristics of diesel engine fumigated with ammonia. In Proceedings of the ASME 2018 Internal Combustion Engine Division Fall Technical Conference, San Diego, CA, USA, 7 November 2018. [CrossRef]
13. Tay, K.L.; Wang, W.M.; Li, J.; Zhou, D. Numerical investigation on the combustion and emissions of a kerosene-diesel fueled compression ignition engine assisted by ammonia fumigation. *Appl. Energy* **2017**, *204*, 1476–1488. [CrossRef]
14. Tay, K.L.; Yang, W.; Chou, S.K.; Zhou, D.; Li, J.; Yu, W.; Zhao, F.; Mohan, B. Effects of injection timing and pilot fuel on the combustion of a kerosene-diesel/ammonia dual fuel engine: A numerical study. *Energy Procedia* **2017**, *105*, 4621–4626. [CrossRef]
15. Kim, K.; Roh, G.; Kim, W.; Chun, K. A preliminary study on an alternative ship propulsion system fueled by ammonia: Environmental and economic assessments. *J. Mar. Sci. Eng.* **2020**, *8*, 183. [CrossRef]
16. Ra, Y.; Reitz, R.D. A reduced chemical kinetic model for IC engine combustion simulations with primary reference fuels. *Combust. Flame* **2008**, *155*, 713–738. [CrossRef]
17. Mathieu, O.; Petersen, E.L. Experimental and modeling study on the high-temperature oxidation of ammonia and related NOx chemistry. *Combust. Flame* **2015**, *162*, 554–570. [CrossRef]
18. Miller, J.A.; Glarborg, P. *Modeling the Formation of N_2O and NO_2 in the Thermal DeNOx Process*; Springer Series in Chemical Physic; Springer: Berlin, Germany, 1996; pp. 318–333.
19. Dukowicz, J.K. A particle-fluid numerical model for liquid sprays. *J. Comput. Phys.* **1980**, *35*, 229–253. [CrossRef]
20. Ricart, L.M.; Xin, J.; Bower, G.R.; Reitz, R.D. *In-Cylinder Measurement and Modeling of Liquid Fuel Spray Penetration in a Heavy-Duty Diesel Engine. SAE Technical Paper 971591*; SAE International: Warrendale, PA, USA, 1997. [CrossRef]
21. Lamas Galdo, M.I.; Rodriguez García, J.D.; Rodriguez Vidal, C.G. Modelo de mecánica de fluidos computacional para el estudio de la combustión en un motor diésel de cuatro tiempos. *DYNA Ing. E Ind.* **2013**, *88*, 91–98. [CrossRef]
22. Lamas, M.I.; Rodriguez, C.G. Numerical model to analyze NOx reduction by ammonia injection in diesel-hydrogen engines. *Int. J. Hydrog. Energy.* **2017**, *42*, 26132–26141. [CrossRef]
23. Lamas, M.I.; Rodriguez, C.G. NOx reduction in diesel-hydrogen engines using different strategies of ammonia injection. *Energies* **2019**, *12*, 1255. [CrossRef]
24. Heiwood, H.B. *Internal Combustion Engine Fundamentals*, 2nd ed.; McGraw-Hill: New York, NY, USA, 1988; ISBN 007028637X.
25. Reusser, C.A.; Pérez Osses, J.R. Challenges for zero-emissions ship. *J. Mar. Sci. Eng.* **2021**, *9*, 1042. [CrossRef]
26. Mallouppas, G.; Yfantis, E.A. Decarbonization in shipping industry: A review of research, technology development, and innovation proposals. *J. Mar. Sci. Eng.* **2021**, *9*, 415. [CrossRef]

27. Lamas, M.I.; Rodriguez, C.G.; Aas, H.P. Computational fluid dynamics analysis of NOx and other pollutants in the MAN B&W 7S50MC marine engine and effect of EGR and water addition. *Trans. RINA Int. J. Marit. Eng.* **2013**, *155*, 81–88.
28. Lamas, M.I.; Rodriguez, C.G.; Rodriguez, J.D.; Telmo, J. Computational fluid dynamics analysis of NOx reduction by ammonia injection in the MAN B&W 7S50MC marine engine. *Trans. RINA Int. J. Marit. Eng.* **2014**, *156*, 213–220. [CrossRef]
29. Lamas, M.I.; Rodríguez, C.G.; Rodriguez, J.D.; Telmo, J. Internal modifications to reduce pollutant emissions from marine engines. A numerical approach. *Int. J. Nav. Archit. Ocean Eng.* **2013**, *5*, 493–501. [CrossRef]
30. Rodriguez, C.G.; Lamas, M.I.; Rodriguez, J.D.; Caccia, C. Analysis of the pre-injection configuration in a marine engine through several MCDM techniques. *Brodogradnja* **2021**, *72*, 1–17. [CrossRef]
31. Rodriguez, C.G.; Lamas, M.I.; Rodriguez, J.D.; Abbas, A. Analysis of the pre-injection system of a marine diesel engine through multiple-criteria decision making and artificial neural networks. *Polish Marit. Res.* **2021**, *28*, 88–96. [CrossRef]

Journal of
Marine Science and Engineering

Article

Assessment of Technical and Ecological Parameters of a Diesel Engine in the Application of New Samples of Biofuels

Juraj Jablonický [1], Patrícia Feriancová [1], Juraj Tulík [1,*], Ľubomír Hujo [1], Zdenko Tkáč [1], Peter Kuchar [2], Milan Tomić [3] and Jerzy Kaszkowiak [4]

[1] Department of Transport and Handling, Faculty of Engineering, Slovak University of Agriculture in Nitra, Tr. A. Hlinku 2, 949 76 Nitra, Slovakia; juraj.jablonicky@uniag.sk (J.J.); xferiancova@uniag.sk (P.F.); lubomir.hujo@uniag.sk (Ľ.H.); zdenko.tkac@uniag.sk (Z.T.)

[2] Department of Building Equipment and Technology Safety, Faculty of Engineering, Slovak University of Agriculture in Nitra, Tr. A. Hlinku 2, 949 76 Nitra, Slovakia; peter.kuchar@uniag.sk

[3] Department of Agricultural Engineering, Faculty of Agriculture, University of Novi Sad, Dr Zorana Đinđića 1, 211 02 Novi Sad, Serbia; milanto@polj.uns.ac.rs

[4] Faculty of Mechanical Engineering, University of Technology and Life Sciences in Bydgoszcz, Al. Prof. S. Kaliskiego 7, 85-789 Bydgoszcz, Poland; kaszk@utp.edu.pl

* Correspondence: juraj.tulik@uniag.sk; Tel.: +421-37-641-4531

Abstract: The technical and environmental parameters of the diesel internal combustion engine using two new samples of biofuels SAMPLE 1 and SAMPLE 2 were evaluated in this paper. SAMPLE 1 and SAMPLE 2 biofuels were tested on a LOMBARDINI LDW 502 internal combustion engine, which was loaded on a dynamometer according to the applicable national and international standards. This method can also be applied to marine engines and contribute to a higher level of marine ecology. The result of the testing was to determine the impact of tested biofuels on the technical parameters engine power and torque and the environmental parameters emissions of smoke, nitrogen oxides, and economy of the internal combustion engine-specific fuel consumption. From the measured data, another parameter was calculated, such as the injected fuel dose and the overall efficiency of the internal combustion engine. The results show that the new samples of SAMPLE 1 and SAMPLE 2 biofuels tested could be a suitable alternative to standard diesel.

Keywords: emissions; alternative fuels; combustion engine; environment

Citation: Jablonický, J.; Feriancová, P.; Tulík, J.; Hujo, Ľ.; Tkáč, Z.; Kuchar, P.; Tomić, M.; Kaszkowiak, J. Assessment of Technical and Ecological Parameters of a Diesel Engine in the Application of New Samples of Biofuels. *J. Mar. Sci. Eng.* **2022**, *10*, 1. https://doi.org/10.3390/jmse10010001

Academic Editor: María Isabel Lamas Galdo

Received: 9 November 2021
Accepted: 19 December 2021
Published: 21 December 2021

Publisher's Note: MDPI stays neutral with regard to jurisdictional claims in published maps and institutional affiliations.

1. Introduction

At present, the issue of environmental studies and energy is one of the serious problems in society [1,2]. Transport has a significant impact on the creation of undesirable emissions as well as energy load. Depending on the level of motorization, the transport sector contributes to air pollution (CO_2) in the range of approximately 11% [3]. The main share of liquid driving fuel falls on road and rail transport. Energy consumption in transport is increasing both in absolute terms, i.e., in energy units, but also in relative terms, which can be characterized as its share in the total energy consumption of all sectors in all major regions of the world [4]. At present, the transport sector contributes circa 32% of the EU's energy consumption. We can therefore state that the transport sector is one of the most demanding in terms of energy consumption [5].

The power unit in mobile energy used in transport is an internal combustion engine. One of the disadvantages of an internal combustion engine is the products of combustion, i.e., exhaust emissions. Exhaust emissions from a diesel engine have a negative impact on the environment as well as on humans [6]. These are mainly emissions of pollutants, especially "greenhouse" gases, which cause gradual irreversible warming of the planet and disequilibriate in nature with acid rain. This is mainly carbon dioxide CO_2, nitrogen oxides NO_X, methane CH_4, and sulfur oxides SO_X [7]. The exhaust gases of a diesel engine form a complex mixture of compounds present in the solid and gaseous phases and include certain classes of compounds, such as polycyclic aromatic hydrocarbons, many of which are genotoxic. These gases are released into the air by burning fossil fuels (crude oil, coal, and natural gas) [8]. The European Community, as the regulatory authority for motor vehicles, has been setting limit requirements for vehicles for more than 20 years in terms of their environmental impact [9]. Arrangements have been proposed to achieve the goal of reducing greenhouse gas emissions, which should replace up to 10% of standard hydrocarbon fuels in biofuels in the near future. Many technologies are currently being used to reduce particulate emissions, including engine constructional modifications, emission control devices, and the use of alternative fuels [10].

The composition of exhaust gases and the proportion of the individual components of the diesel engine are relatively balanced [11]. During the technical life of the engine, there are no gradual changes in the difference in the proportions of individual components in all modes of engine operation. A change in the technical condition of the engine affecting the quality of the mixture preparation or the conditions of the combustion process causes an increase in the production of CO resp. HC. This is reflected, on the one hand, in a decrease in the effective pressure, and thus in an improvement in the conditions for reducing the content of nitrogen oxides NO_X, and, on the other hand, a change in the temperature balance has the effect of increasing the production of solid particles. It is clear from this that the deterioration of the technical condition of the engine while maintaining the transport characteristics of the injection pump will be reflected in a decrease in the engine torque while maintaining hourly fuel consumption and thus an increase in specific fuel consumption. This condition is reflected in a change in the shares of individual pollutants in the exhaust emission, namely by a decrease in the proportion of NO_X and an increase in the proportions of CO and HC with a simultaneous increase in the smoke of the engine [12].

The possibility of creating new emissions should also be considered if additives are applied to the fuel or lubricating oil and where fluids are introduced into the exhaust gases. A well-known example is urea used to reduce NO_X emissions in an SCR catalyst. SCR can include ammonia, as well as a number of substances from incomplete decomposition of urea [13].

One way to comply with stricter emissions regulations is to focus attention and look for suitable alternative fuels as reported by [14] in his publication. According to these authors, the main alternative fuels used in automobile transport could be ethanol, hydrogen, and biodiesel. A large number of studies have shown that biodiesel could be an alternative fuel, which is one of the fuels that could be used in diesel engines with little or no treatment requirements with little or no adjustment requirements [15]. It has been proven that biodiesel has significant potential for reducing CO_2, CO, THC, and PM emissions [16,17]. Studies on biodiesel have been extensively researched due to their attributes of ecological and economic viability and their aspects [18]. Biodiesel consists of various organic substances. These substances can be combined with fuel in various ratios [19]. It is not necessary to modify the diesel engine to use biodiesel, because the properties of biodiesel are similar to conventional fuels. The advantage of biodiesel is that it is a renewable energy source, biodegradable, environmentally friendly, non-explosive, and produces less emissions [20]. The main components of biodiesel are cooking oil, vegetable waste and oils, and animal fats. The use of these oils and fats is inconvenient without further treatment because they have a higher viscosity than normal crude oil [21]. Biofuel sources are divided into edible (sunflower oil, peanut oil, coconut oil, palm oil, etc.) and inedible (jojoba, jatropha, cotton, moringa, etc.) sources [22]. Studies have shown that a mixture of biodiesel reduces emissions of gases (CO, PM, and HC) that are toxic to the environment and increase air pollution [21]. When comparing biofuels with conventional oil, both substances have the same physicochemical properties. Thanks to these properties, the engine needs only a small modification [23]. Mixing diesel with inedible oil increases the given properties of the fuel. Antony et al. studied the combination of fuel with candle and mind oil. This combination showed reduced CO, CO_2, HC, and NO_X emissions compared to pure diesel [24]. The authors Sharma et al., researched the energy and emission parameters of the hydroxyl diesel engine. These results revealed that CO, unburned hydrocarbon, and smoke emissions were reduced, and NO_X emission enhanced [25]. The authors Sarıkoç et al. investigated the performance of the diesel engine with different biodiesel and diesel-butanol blends. These results revealed that biodiesel increases the energy efficiency and that 75% diesel, 20% biodiesel, and 5% butanol is the best fuel ratio for sustainability parameters [26]. The authors Gad et al. performed an analysis of a diesel engine with waste cooking oil biodiesel and gasoline. These results revealed that the addition of gasoline reduces the opacity of CO, NO_X, unburned hydrocarbon, and smoke and that a mixture of 92% waste cooking oil and 8% gasoline is the best combination in terms of the performance and emission characteristics [27]. Alternative fuels made from agricultural products have a higher oxygen content compared to diesel. The authors Ramadhas et al. state that the use of mixtures of methyl esters of vegetable oils achieves the same engine power as diesel at full load conditions, as well as the average load at different engine speeds [28]. The authors Qi et al. and Frijters & Baert agree that the lower heat value of biodiesel must be offset by its higher consumption. The performance of an internal combustion engine is affected by the properties of the biofuel, in particular its heat value, viscosity, and lubricating properties [29,30].

It can be stated that alternative fuels or their mixtures formed with diesel are constantly the subject of research, with a focus on reducing emissions arising from their combustion in the engine, but also the conversion of their heat energy into mechanical work [31]. Currently, researchers are involved in environmental protection and the production of biofuels and its use in various equipment [32]. Many research articles on energy, emission, and combustion analysis of biodiesel engines are available, such as [33,34].

2. Materials and Methods

2.1. Test Conditions for Engines in Laboratory Conditions

According to ISO 2288, STN 30 0415, and STN ISO 789–1 (30 0411), the conditions for testing tractor engines in laboratory conditions are as follows:

- The test must be performed at the set full fuel dose on the injection pump;
- The test shall be performed on the engine after running-in, as consistent with the manufacturer's recommendations, under conditions of stable engine operation;
- When preparing the engine for the tests, a setting must be made, which must not be changed during the tests;
- Engines supplied for testing must be equipped completely according to the manufacturer's enclosed technical documentation, with documents of acceptance by the technical inspection of the production company and the running-in protocol;
- All repair principles and deficiencies must be recorded during the tests;
- The atmospheric conditions of the test should be as close as possible to the reference to reduce the value of the correction factor. This represents an atmospheric pressure of 96.6 kPa and a temperature of $23 \pm 7\,°C$;
- The temperature of the air entering the engine must be measured at a maximum distance of 0.15 m from the inlet. The temperature sensor should be protected from radiant heat and placed in the air flow;
- The countdown can only take place when the torque, speed, and temperature are approximately constant for at least one minute;
- The speed value after setting it must not change $\Delta n = \pm 1\%$ or $10\ \mathrm{min}^{-1}$, taking into account the higher value;
- The average value of two stable measurements (differing by a maximum of 2%) of load, consumption, and intake air temperature shall be recorded;
- For temperature and speed measurements, a minimum measurement time of 30 s is set for automatic measurement and 50 s for manual measurement;
- Unless otherwise specified by the manufacturer, the outlet coolant temperature shall be $80 \pm 5\,°C$ and the oil temperature shall be between 80 and $100\,°C$ for air-cooled

engines, and the temperature must be maintained at the exact location specified by the manufacturer with a tolerance of $\pm 20\ ^\circ C$;

- The fuel temperatures at the inlet to the injection pump and crankcase (or at the outlet from the oil cooler) must be within $\Delta t \pm 5\ ^\circ C$ of the temperature specified by the engine manufacturer with a minimum value of $30\ ^\circ C$;
- The exhaust outlet temperature must be measured at the exhaust flange and must be within the specified range;
- The exhaust system in the test cell must not, during engine operation, generate a pressure other than atmospheric pressure with a tolerance of $\Delta p \pm 740$ Pa (measured at the connection point);
- The cooler, blower, water pump, and thermostat must be located as mounted on the vehicle during the test;
- Auxiliary equipment is specified, which must be included in the test; others must be excluded during the test;
- Engine tests should be performed without interruption. When it is necessary to shut down the engine during the tests, it must be ensured that it warms up to the prescribed steady values [35].

The accelerator pedal was pressed quickly but non-violently (at most in 1 s) to achieve the maximum fuel quantity, and it was released until the maximum rpm had been reached and recorded by the device, i.e., until approximately 2 s after reaching them. After releasing the accelerator pedal, a decrease in speed was expected, which could not be higher than the value of idling speed. Subsequently, the values of the absorption coefficient, idle rpm, maximum speed, and acceleration time were recorded. Free acceleration was performed 3 times, and the time between two successive presses of the accelerator pedal was at least 10 s. A flow meter from EMICRON DMS 1 was used to measure consumption, which measures with an accuracy of up to 1%. The accuracy of the dyno is 2%.

2.2. Characteristics of the Measured Object

An Italian-made LOMBARDINI LDW 502 diesel engine, which meets the EURO 2 emission standard, was used for the measurement. The LOMBARDINI LDW 502 engine is a water-cooled four-stroke in-line twin-cylinder with direct fuel injection and a pump-nozzle injection system. The engine distribution is an overhead valve and there are two valves per cylinder. The engine is not supercharged. The basic parameters of engine are in Table 1.

Table 1. Basic technical parameters of the LOMBARDINI LDW 502 engine.

Parameter Name	Values
Number of engine cylinders, pcs	2
Engine capacity, cm^3	505
Cylinder diameter, mm	72
Stroke, mm	62
Compression ratio, –	22.8:1
Combustion order, $^\circ$	360
Engine power (according to 80/1269/CEE), ISO 1585, kW/HP	4.0/5.43
Maximum torque, Nm	20
Rpm at maximum torque, min^{-1}	1600
Engine idle, min^{-1}	950–1100
Nominal rpm, min^{-1}	3000
Maximum rpm, min^{-1}	3400
Water pump flow at 3400 min^{-1}, dm^3	40
Air consumption at 3400 min^{-1}, dm^3	910

The emission state of the combustion engine was measured with an exhaust gas analyzer MET 6.3 from MAHA. The meter operates with an expanded measurement uncertainty U, which is expressed as the standard measurement uncertainty multiplied by the overlap coefficient $k = 2$, which, under a normal distribution, corresponds to a probability of coverage of approximately 95% and was set according to MSA 0104-97, respectively EA-4/02 M:2013.

The basic technical data of the emission analyzer MET 6.3 are listed in Table 2.

Table 2. Technical data of the emission analyzer MET 6.3.

Measurable Gases	HC, CO, CO$_2$, O$_2$, NO, NO$_2$, NOx
Flow	3.5 dm^3 min^{-1}
Accuracy class	O (OIML)
Measurement principle	Absorbance measurement
Warm-up time of the measurement chamber approx.	150 s
Measuring range particle concentration	1–1100 mg·m^{-3}
Resolution of particle concentration	1
Measurement range turbidity	0–100%
Measuring range absorption coefficient	0–9.99 m^{-1}
Resolution absorption coefficient	9.99 m^{-1}
On-board voltage	10 V/30 V
Power supply	1/N/PE 110 V/230 V 50 Hz/60 Hz
Ambient temperature	0–45 °C

The aim was to assess new samples of biofuels with a focus on the technical-operational parameters, emission parameters, and fuel consumption in accordance with the requirements of the manufacturer of alternative fuels by the designed laboratory device (Figure 1).

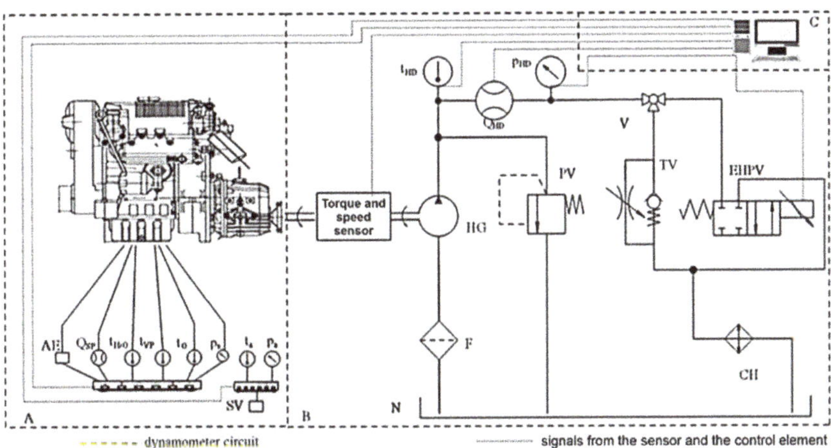

Figure 1. Experimental installation. A—internal combustion engine connection circuit, B—hydrostatic dynamometer circuit, C—measuring and evaluation circuit, N—tank, F—filter, HG—hydrogenerator, t_{HD}—temperature sensor, Q_{HD}—flow meter, p_{HD}—pressure gauge, PV—safety valve, V—three-way valve, TV—throttle valve, EHPV—electrohydrostatic proportional valve, CH—cooler, AE—emission analyzer, Q_{SP}—fuel consumption meter—EMICRON DMS, t_{H_2O}—water temperature, t_{VP}—exhaust gas temperature, t_O—oil temperature, p_s—intake manifold pressure, t_a—ambient temperature, p_a—atmospheric pressure, SV—humidity sensor.

Characteristics of the Fuels Used to Monitor the Emission Status of the Combustion Engine

Samples of biofuels were provided for experimental purposes and assessment in accordance with the manufacturer. However, the manufacturer provided only the basic parameters of the samples. For this reason, only the basic fuel specification is given in the tables.

During the measurements, diesel was used in the fuel system, the basic specifications of which are listed in Table 3.

Table 3. Diesel specification according to the standard EN 590.

Fuel Properties	Unit	Specification		Test
		Min	**Max**	
Cetane number		49	–	ISO 5165
Cetane index		46	–	ISO 4264
Density at 15 °C	$kg \cdot m^{-3}$	820	860	ISO 3675/ASTM D4052
Sulfur content	% (weight)	–	0.20	EN 24260/ISO 8754
Flash point	°C	55	–	ISO 2719
Carbon residues (10% btms)	% (weight)	–	0.30	ISO 10370
Ash	% (weight)	–	0.01	EN 26245
Water content	$mg \cdot kg^{-1}$	–	200	ASTM D1744
Copper strip corrosion, 3 h at 50 °C		–	Class I	ISO 2160
Oxidation stability	$g \cdot m^{-3}$	–	25	ASTM D2247
Viscosity at 40 °C	$mm^2 \cdot s^{-1}$	1.9	6.0	ASTM D 6751
Distillation 10% point 50% point 65% point	°C	250	–	ISO 3405
85% point		–	350	
95% point		–	370	

The results of technical-exploitative and emission tests were compared with the results, which were achieved by measuring the reference sample, which is diesel.

Tables 4 and 5 show the parameters of the alternative fuel SAMPLE 1 and SAMPLE 2, the new sample.

Table 4. Properties of alternative fuel SAMPLE 1.

Properties	Unit	Determined Value
Heat value	$kJ \cdot kg^{-1}$	39,390
Density at 15 °C	$kg \cdot m^{-3}$	880
Kinematic viscosity at 40 °C	$mm^2 \cdot s^{-1}$	5.5
Flash point	°C	130
Pour point	°C	−20
Filterability	°C	−15
Sulfur content max.	% hm. max.	0.05
Cetane number	–	55
Ash content	% hm. max.	0.01
Water content	% hm. max.	0.05
Acid number	$mg \cdot KOH^{-1}$ max.	0.3
Content of mechanical impurities	% hm. max.	0.005

Table 5. Properties of alternative fuel SAMPLE 2.

Properties	Unit	Determined Value
Heat value	$kJ \cdot kg^{-1}$	40,763
Density at 15 °C	$kg \cdot m^{-3}$	880
Kinematic viscosity at 40 °C	$mm^2 \cdot s^{-1}$	4.2
Flash point	°C	145
Pour point	°C	−17
Filterability	°C	−21
Sulfur content max.	% hm. max.	9.08
Cetane number	–	55.8
Ash content	% hm. max.	0.0035
Water content	% hm. max.	338
Acid number	$mg \cdot KOH^{-1}$ max.	0.41
Content of mechanical impurities	% hm. max.	5
Total glycerol	% hm.	0.2
Methyl ester content	% hm.	97
Phosphorus content	$mg \cdot kg^{-1}$	1.01

The test methods for determining the parameters listed in Tables 4 and 5 were determined according to the following test methods: STN EN ISO 12937, STN EN 14105, STN EN ISO 12185, STN EN ISO 20846, STN EN ISO 5165, ASTM D 3231, ASTM D 6751, STN 65 6063, STN EN ISO 3104, STN EN ISO 2719, STN EN 116, STN EN ISO 10370, STN EN 14104, STN EN 12662, STN EN 14103.

2.3. Calculation of the Correction Factor K_a and the Motor Factor f_m

When determining the correction factor K_a, the absolute temperature (T) of the air drawn into the engine, expressed in kelvins, and the dry atmospheric pressure (p) expressed in kPa or mbar, and according to the relations below, the parameter K_a is determined.

Calculation of the Correction Factor for Diesel Engines without Supercharging or with Mechanical Supercharging

To calculate the correction factor for diesel engines without supercharging respectively with mechanical supercharging, relations according to DIN, EWG, ISO, SAE, and JIS standards are used. For standards EWG 80/1269, ISO 1585, SAE J1349, and JIS D1001, the calculation of the correction factor according to the relation:

$$K_a = \left[\frac{990}{p} \cdot \left(\frac{T}{298} \right)^{0.7} \right]^{fm} \tag{1}$$

The following applies to DIN 70020:

$$K_a = \frac{1013}{p} \cdot \left(\frac{T}{293} \right)^{0.5} \tag{2}$$

2.4. Uncertainty Analysis

The measurement uncertainty for the MET6.3 emission analyzer is $U = 2$.

Mathematical Statistical Analysis of the Processing Achieved Results

The determination of the required number of repetitions of the experiment can be performed according to the given methods if the standard deviation of the measured quantity or the coefficient of variation is known in advance. These values can be known from previous measurements, preliminary control measurements, or the literature. The determination of these basic parameters of descriptive statistics is given in the following section. If the standard deviation is known, then the number of repetitions of the experiment can be determined for the selection with repetition according to Equation (3).

To determine the probability value, i.e., assumption of the occurrence of analogous results when repeating the experiment, there are no precise rules. In experiments associated with the construction of machines, a probability of 90% is required. The relationship for determining the number of repetitions can be written as follows:

$$n = t_a^2 \cdot \frac{s}{\Delta^2} \tag{3}$$

where:
    n is the number of repetitions detected;
    t_a is the critical value of the normal distribution for a bilateral estimate;
    Δ is the selected value of the required accuracy;
    S is the estimate of the standard deviation of the base set.

In experiments in which partial regularities and values of quantities serving as a basis for further calculations are determined, it is necessary that the probability achieves 99% [34].

The following relationship can be used to determine the required number of set repeats [13]:

$$n = \frac{N \cdot t_a^2 \cdot s^2}{t_a^2 \cdot s^2 + N \cdot \Delta} \tag{4}$$

where:

N is the extent of the set used.

This calculation method can also be used for the base set, because at values $N > 10^6$, the required number of repetitions n further increases very slightly.

If the coefficient of variation of the set v_k is known, then the required number of repetitions can be determined according to the relation:

$$n = \frac{v_k^2 \cdot t_a^2}{\delta^2} \tag{5}$$

where:

n is the number of repetitions detected;

t_a is the critical value of the normal distribution for a bilateral estimate;

v_k is the coefficient of variation, %;

δ is the maximum permissible error, %.

For a known value of the coefficient of variation, it is possible to determine the required number of repetitions using the ratio δ/v_k.

As far as the measurement of one quantity is concerned, the task of the proposal is reduced to determine the number of repeated measurements. For repeated measurements under the same conditions, the number of measurements is determined either by the prescribed uncertainty of the measurement result or by the prescribed costs. The uncertainty of the measurement result is formed by the uncertainty evaluated by the type A method and the uncertainty evaluated by the type B method. The first component of uncertainty is affected by the number of measurements and the second component by the measurement method, measuring equipment, measurement conditions, etc. In order to achieve the expected uncertainty of the measurement result, we had to choose the appropriate method, measuring equipment, and measurement conditions. By the number of measurements, we were able to influence only the component of uncertainty determined by the type A method [35].

For repeated measurement under the same conditions for the measured values X_1, X_2, ..., X_n, the estimate of the value of the measured quantity is given by the arithmetic mean of the measured values (at zero corrections) and the uncertainty assessed by the type A method by:

$$\mu_A = \frac{s}{\sqrt{n}} \tag{6}$$

where:

s is the selection standard deviation.

The sample standard deviation of the data is defined as the square root of the sample variance. We calculated the sample standard deviation from the realized selection, i.e., if we do not have a complete set of possible states but only a selection of them. For example, we measured the value of a certain physical quantity and repeated the measurement, e.g., 10 times. Because each equipment has its own prescribed accuracy class, the results of our measurements varied slightly at the lowest order locations. Then, we cannot determine the standard deviation for a small number of measurements. We must therefore consider $n-1$ in the formula (because 1 measurement is already dependent on the calculation of the mean value). For a large number of measurements, the difference between the standard and sample standard deviation is lost [36].

The sample standard deviation is then given by:

$$s = \sqrt{\frac{1}{n-1} \sum_{i=1}^{n} (x_i - \overline{x})^2}$$ (7)

where:

x_i is the measured values;

$\overline{x}$ is the arithmetic mean.

Equation (7) results in the recalculation of repeated measures characterized by the variance s^2 to achieve a given uncertainty component μ_A:

$$n \geq \left(\frac{s}{\mu_A}\right)^2$$ (8)

This means that if we want to reduce the uncertainty component of type A by k times, we must increase the number of measurements k by 2 times with constant variance s^2. Another way is to take measures to reduce the variance of measurements.

When measuring several quantities, we want to find a measurement proposal for which the uncertainties in determining the values of all quantities would be the smallest of all possible proposals. We call such a proposal evenly the best proposal. In reality, however, the existence of a uniformly best proposal is rare. Therefore, we must most often specify the goal of the experiment (measurements), and choose the appropriate proposal selection criterion [36].

2.5. Influence of the Tested Fuel on the Technical and Ecological Parameters of the LOMBARDINI LDW 502 Engine

The technical and ecological parameters of the LOMBARDINI LDW 502 engine were determined during the combustion of standard hydrocarbon fuel and two samples of alternative fuels. The engine is made by Lombradini a Kohler Company, Emilia, Italy. Diesel was the reference fuel on the basis of which the data obtained during the application and combustion of the alternative fuel SAMPLE 1 and SAMPLE 2 were compared with a new sample in the fuel system of the above engine.

The benchmarks for all fuels tested included:

- Torque M_k, Nm;
- Corrected power P_{ekor}, kW;
- Injected fuel dose, $\mu L \cdot cylinder^{-1} \cdot cycle^{-1}$;
- Specific fuel consumption m_{pe}, $g \cdot kW^{-1} \cdot h^{-1}$;
- Thermal efficiency η_t, %;
- Smoke value, k, m^{-1};
- NO_X emission value, ppm.

To determine the thermal efficiency of the engine η_t, it is necessary to know the energy, which is contained in the fuel E_p, which can be expressed by the relation:

$$E_p = \frac{M_p \cdot HU}{3.6} \quad kJ \cdot s^{-1}$$ (9)

where:

M_p is the hourly fuel consumption, $kg \cdot h^{-1}$;

HU is the lower heat value of fuel, $MJ \cdot kg^{-1}$.

$$\eta_t = \frac{P_{ekor}}{E_p} \cdot 100\%$$ (10)

Measurement of Technical and Ecological Parameters on the Test Engine for Discrete Measurement

Results of measurement of technical and ecological parameters of the engine on a test rig at steady state with fuel: diesel.

Conditions under which the measurements were made:

Laboratory temperature	24.2 °C = 297.35 K
Atmospheric pressure	99.17 kPa
Humidity	30.90%
Fuel density–diesel	0.84 g·cm^{-3}
Fuel heat value–diesel	41.90 MJ·kg^{-1}

Results of measurement of technical and ecological parameters of the engine on a test rig at steady state with fuel: SAMPLE 1:

Conditions under which the measurements were made:

Laboratory temperature	23.7 °C = 296.85 K
Atmospheric pressure	101.3 kPa
Humidity	26.57%
Fuel density–SAMPLE 1	0.88 g·cm^{-3}
Fuel heat value–SAMPLE 1	38.50 MJ·kg^{-1}

Results of measurement of technical and ecological parameters of the engine on a test rig at steady state with fuel: SAMPLE 2:

Conditions under which the measurements were made:

Laboratory temperature	24 °C = 297.15 K
Atmospheric pressure	101.5 kPa
Humidity	31.27%
Fuel density–SAMPLE 2–new sample	0.88 g·cm^{-3}
Fuel heat value–SAMPLE 2–new sample	38.00 MJ·kg^{-1}

3. Results

Evaluation of Measured Data from Tests of Tested Fuels on a Test Device at Steady State Mode with the LOMBARDINI LDW 502 Test Engine

Statistically, the minimum number of measurements was calculated to be three. We performed seven measurements for each tested fuel. The correction factor was calculated according to Formula (1), valid for EGG 80/1269, which is used for non-supercharged engines. If the motor has no engine factor, the parameter is specified according to the engine type, the engine factor is set at 0.3 also in accordance with the manufacturer's recommendations. The operating conditions for the tests were as follows:

The engine temperature during the measurement ranged from 80 to 92 °C, and the speed range during loading was from 1200 to 3200 min^{-1}. The tests were performed according to ISO 2288, STN 30 0415, and STN ISO 789-1 (30 0411).

Based on the measured data of the tested fuels, the graphical dependences of the compared samples of biofuels with the reference fuel were constructed. The results of the measurements are recorded in graphical form in Figures 2–6.

Figure 2 shows the relationship of torque and fuel injection to rpm for diesel fuels SAMPLE 1 and SAMPLE 2. It can be seen from the graphic expression that there are minimal differences between the individual fuels, both during the torque curves and in the achieved torque values. The torque values for SAMPLE 1 fuel were increased by 2.4% at 3200 rpm compared to diesel fuel; however, in the speed range at maximum torque, a decrease of 0.3% was recorded, which is a negligible value. It is worth mentioning the recorded value of torque at the speed of 2800 rpm and the course of the torque curve from the speed of 1600 rpm to the speed of 1200 rpm for the mentioned tested fuel. The increase in torque at 2800 rpm as a linear course of torque in the speed range from 1600 to 1200 rpm is due to the amount of fuel injected per cycle, which is directly proportional to the hourly fuel consumption M_p and inversely proportional to the engine speed n and the density of the fuel ρ.

When comparing SAMPLE 2 fuel with the reference diesel fuel (Figure 2), minimal differences were found during the torque, as well as in the achieved values. The range of torque reduction for SAMPLE 2 fuel is from 0.33 to 2% compared to diesel.

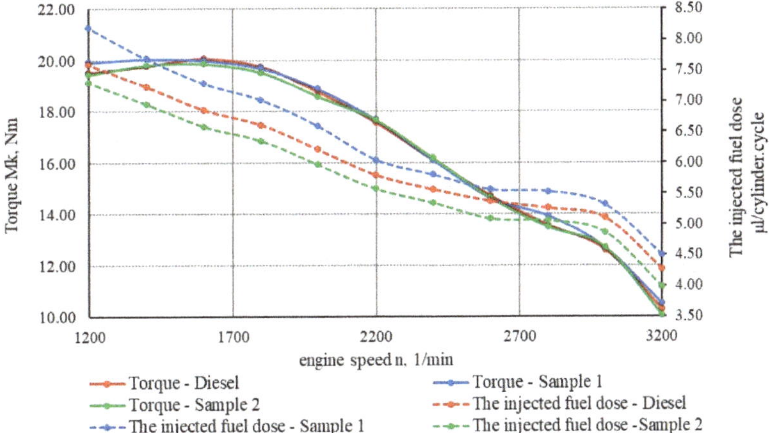

Figure 2. Comparison of the course of torque and injected fuel dose with tested fuels on the LOMBARDINI LDW 502 engine.

Regarding the comparison of the injected fuel dose per cylinder and cycle (Figure 2), for SAMPLE1, the value increased from 3% to 7% compared to the reference diesel. In contrast, with SAMPLE 2 fuel, there was a reduction in the fuel injection in the range of 4% to 7% compared to diesel.

As indicated above, the amount of fuel injected per cylinder and cycle is significantly affected by the bulk density of the fuel. which is also confirmed by the relationship for the calculation of the injected fuel dose (11):

$$V_v = \frac{10^6 \cdot M_p}{2.60 \cdot n \cdot \rho} \; \mu L/\text{cylinder} \cdot \text{cycle} \tag{11}$$

where:

V_v is the fuel injection, $\mu L/\text{cylinder} \cdot \text{cycle}$;

P is the fuel density, $\text{kg} \cdot \text{dm}^{-3}$;

M_p is the hourly fuel consumption, $\text{kg} \cdot \text{h}^{-1}$.

The interdependencies between the rpm, torque, and injected dose for the tested fuels in a 3D view on the LOMBARDINI LDW 502 engine are shown in Figure 3.

An evaluation of the tested biofuels was also performed on the basis of a mutual comparison of the achieved effective power P_e and energy supplied in the fuel E_p. The results of the mutual comparison are shown in graphical form in Figure 4. From the external speed characteristic of the LOMBARDINI LDW 502 engine (Figure 4), it can be seen that the effective engine power is constant in the speed range from 2000 to 3000 rpm.

When comparing the achieved power on the LOMBARDINI LDW 502 engine when applying SAMPLE 1 fuel, it can be stated that in the rpm range at which the above-mentioned diesel engine achieves constant values of effective power, there was a decrease in power ranging from 1.76% to 2.84% compared to diesel. In the speed range from 1200 to 2000 rpm, a decrease in power from 0.5% to 2.5% was recorded.

A decrease in effective power was also observed when SAMPLE 2 fuel was applied. In the speed range from 2000 to 3000 rpm, a decrease in effective power in the range from 2% to 3% was recorded. At 3200 rpm, power decreased by approximately 5% and in the speed range from 1200 to 2000 rpm, a decrease in power from 3% to 4% was recorded.

When comparing the energy supplied in the fuel E_p (Figure 4), it can be seen that the highest value is reached by the fuel curve–SAMPLE 1. Compared to diesel fuel, the curve of energy supplied in the fuel with the above-mentioned alternative fuel is higher in the range from 2% to 6%. A lower decrease in the compared indicator was recorded for fuel SAMPLE 2. For this fuel, a decrease in the energy supplied in the fuel in the range from 2% to 6% was recorded compared to diesel. The indicator 'energy supplied in fuel E_p' is affected by the hourly fuel consumption and the lower heat value of the fuel, as expressed by the relationship (9).

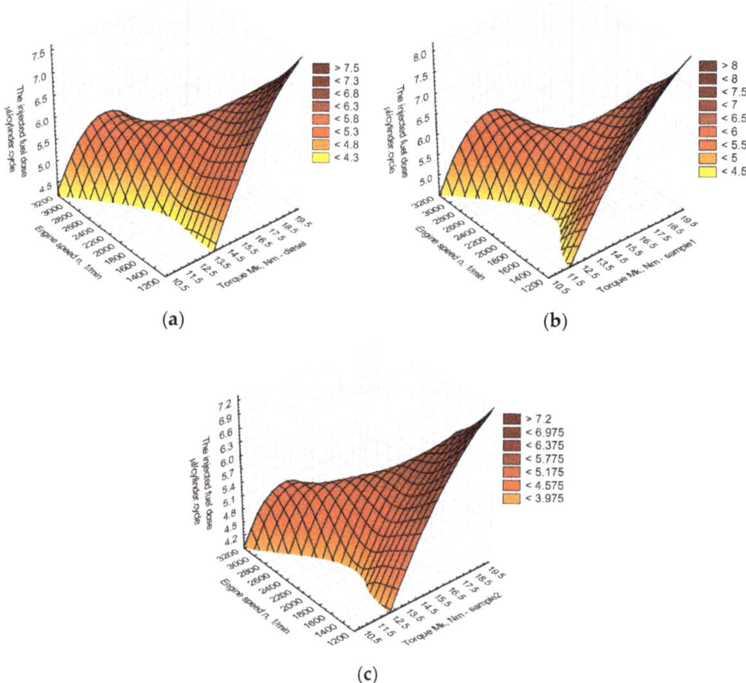

Figure 3. Interdependence of rpm, torque, and injection. (**a**) DIESEL, (**b**) SAMPLE 1, (**c**) SAMPLE 2.

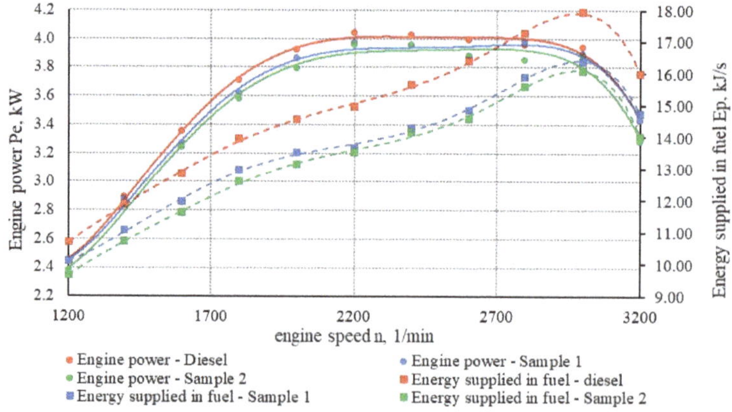

Figure 4. Comparison of the course of power and energy delivered in the fuel with the tested fuels on the LOMBARDINI LDW 502 engine.

The interdependencies between the rpm, torque, and injected dose in the 3D view for the fuels tested on the LOMBARDINI LDW 502 engine are shown in Figure 5.

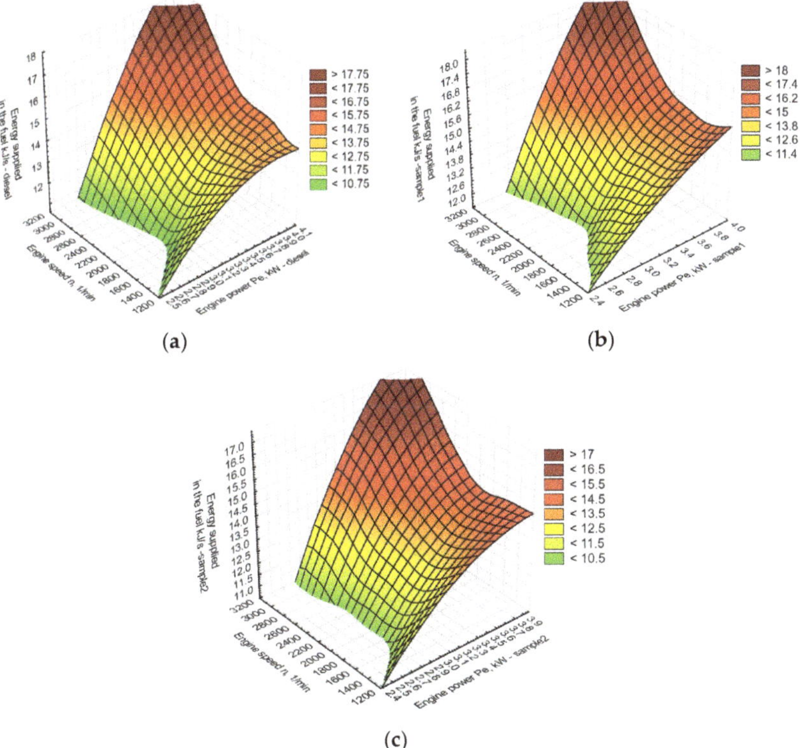

Figure 5. Interdependence of rpm, power, and energy supplied in the fuel. (**a**) DIESEL, (**b**) SAMPLE 1, (**c**) SAMPLE 2.

Other indicators that were evaluated from the point of view of biofuel testing are the specific fuel consumption m_{pe} and the thermal efficiency of the CI engine η_c (Figure 6).

The reference diesel fuel reached the lowest value of specific fuel consumption compared to the compared samples of biofuels. In the SAMPLE 1 sample, the value of specific fuel consumption was higher, in the range from 9% to 12%. The highest difference in specific consumption was recorded in the speed range from 1200 to 2000 rpm.

The values of specific consumption with SAMPLE 2 fuel were higher compared to diesel, in the range from 0.7% to 4%. The highest difference in specific consumption was recorded as in the previous case in the speed range from 1200 to 2000 rpm.

When comparing the tested fuels from the point of view of the thermal efficiency of the LOMBARDINI LDW 502 engine, it can be stated that the highest efficiency was achieved by the diesel engine at 2200 rpm at all tested fuels. The results were calculated from the measured values according to Formulae (9) and (10). At the indicated rpm, the combustion engine operates with the lowest specific fuel consumption. The highest thermal efficiency at 2200 rpm was achieved by the combustion engine with diesel fuel. In the SAMPLE 2 fuel tests, a decrease in thermal efficiency of only 1.4% was recorded at 2200 rpm. Compared to diesel, the value of thermal efficiency decreased from 27% to 26.6%.

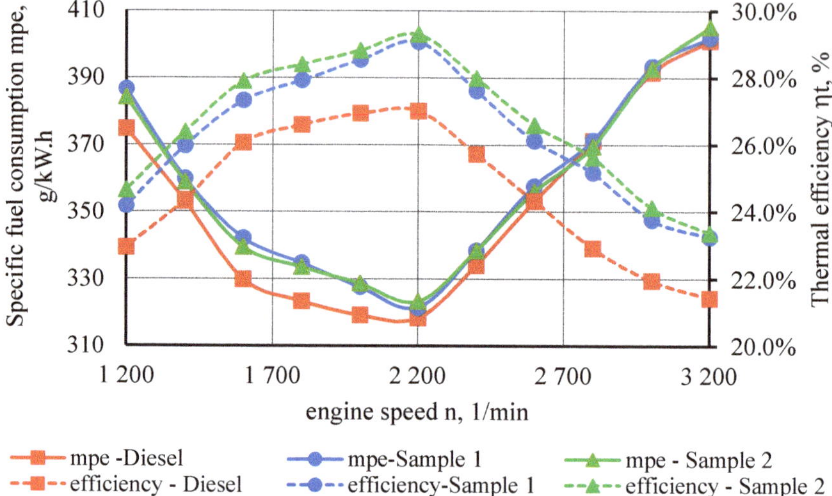

Figure 6. Comparison of the course of specific fuel consumption and thermal efficiency with tested fuels on the LOMBARDINI LDW 502 engine.

When SAMPLE 1 was applied, the value of the thermal efficiency compared to diesel was reduced by 5.6% at 2200 rpm.

The course, as well as the size of the thermal efficiency of the diesel engine is given by the ratio of effective power and energy supplied in the fuel, as given by the relation η_c, which is given in Section 2.5.

From the measured smoke data, it is clear that the range of the MAHA MDO–2 Lon Euro did not allow measurement of the smoke values for SAMPLE 1 and SAMPLE 2 fuels at individual load points of the diesel engine. For this reason, NO_X emission values were monitored and recorded as the LOMBARDINI LDW 502 diesel engine was gradually loaded. A comparison of the measured NO_X emissions from diesel and two alternative fuels is shown in Figure 7.

It was not possible to assess the emission state from the point of view of the absorption coefficient on the tested internal combustion engine. The reason was the low values of the absorption coefficient in the tested samples, which the measuring device could not detect.

A situation in which the analyzer is unable to evaluate the absorption coefficient values also occurs in vehicles equipped with after-treatment devices (EURO 5, EURO 6). Due to the impossibility of assessing the emission state of the diesel engine by measuring the absorption coefficient caused by exhaust gas filtration, it was necessary to find another alternative method of assessing the emission state of the diesel engine. This study focused on NOx emissions in accordance with the requirements of the biofuel sample manufacturer. In the future, NOx emissions may be used to assess the emission status of an internal combustion engine together with methods for detecting malfunctioning, damaged, or dismantled exhaust after-treatment devices.

As can be seen from the graphical comparison of the measured emissions of nitrogen oxides NO_X (Figure 7) under the load of the diesel engine, it can be stated that nitrogen oxides are higher for both tested alternative fuels than for the reference fuel. The average value of nitrogen oxides recorded when loading the LOMBARDINI LDW 502 powered by diesel reached 212 ppm. The measured value is 4.5% lower than the average value of NO_X emissions with the alternative fuel SAMPLE 1 and 7.42% lower than the average NO_X emissions of the alternative fuel SAMPLE 2 (Figure 8).

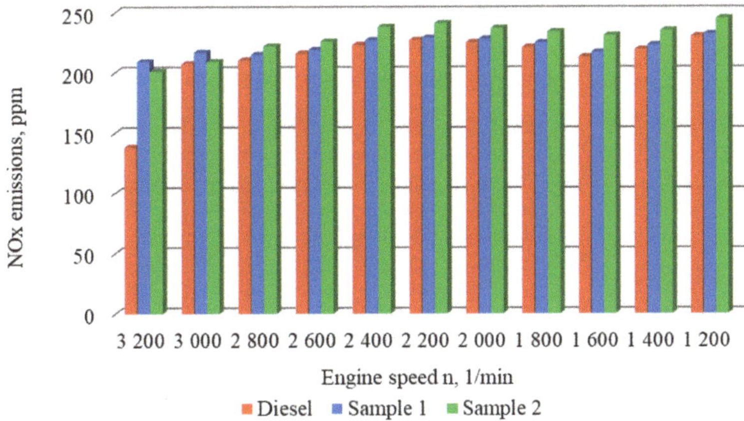

Figure 7. Comparison of NO$_X$ emissions with tested fuels on the engine LOMBARDINI LDW 502 under its load.

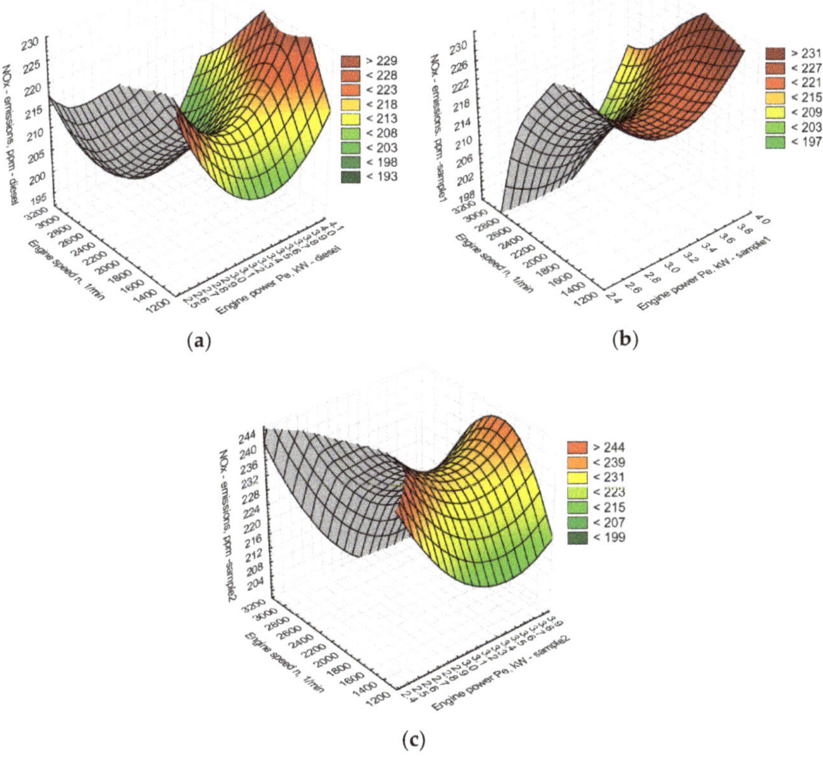

(a)

(b)

(c)

Figure 8. Interdependence of speed, corrected power, and NO$_X$ emissions. (**a**) DIESEL, (**b**) SAMPLE 1, (**c**) SAMPLE 2.

4. Discussion

In the submitted study, two types of alternative fuels, SAMPLE 1 and SAMPLE 2, were compared with diesel as a reference fuel. The comparison of the above-mentioned alternative fuels was intended to monitor their impact on technical parameters, environmental parameters, and economy of the internal combustion engine.

The significant impact regarding the technical parameters of the internal combustion engine when using biofuel as an energy source will be its physicochemical properties. The throughput of an internal combustion engine is affected by the properties of biofuel, in particular its caloricity, viscosity, and lubricating properties. The calorific value of fuels is an important indicator, which affects the release of energy for labor production. The lower calorific value of biodiesel is related to the decrease in engine throughput.

The calorific value of fuels is in an important indicator that affects energy release for labor production. A lower calorific value of biodiesel is related to a decrease in engine throughput. This finding can be confirmed with the test results for fuels SAMPLE 1 and SAMPLE 2, which compared to diesel fuel had a lower calorific value, as well as lower achieved throughput. As for viscosity, we can agree with [14,36], who reported that higher viscosity is the cause of throughput loss, as higher viscosity reduces combustion efficiency due to impaired fuel injection atomization. The authors' opinion can be confirmed given that the tested fuels SAMPLE 1 and SAMPLE 2 showed a higher viscosity value than diesel and, as stated earlier, the conducted measurements confirmed a decrease in throughput.

Specific fuel consumption was the next parameter that was monitored when comparing the fuels SAMPLE 1 and SAMPLE 2. From the achieved results of specific fuel consumption, it can be stated that both tested fuels had a higher consumption rate compared to diesel. Specifically, the SAMPLE 1 fuel recorded an increase in specific fuel consumption by an average of 10.6% and the SAMPLE 2 fuel by 2.6%. The increase in the specific consumption rate of the tested fuels is caused by their lower caloricity and their higher specific weight compared to diesel. Comparable results were also achieved by [37,38], who agrees with the notion that a lower calorific value of biodiesel must be offset by a higher consumption rate.

While testing alternative fuels, the measurement of NO_X emissions was also conducted and focused on. Emissions of nitrogen oxides are significantly influenced by the cetane number of the fuel, the amount of oxygen contained in the fuel, raw material used in the production of the biofuel, and lastly by the injection advance value. As previously mentioned, the cetane number has an effect on the amount of nitrogen oxide NO_X emissions. The tested alternative fuels SAMPLE 1 and SAMPLE 2 have higher values of the cetane number compared to diesel. A higher cetane number results in shortening of the ignition delay, which leads to better combustion but at the same time increases the production of nitrogen oxide NO_X emissions. This statement is also confirmed in the papers [39–44]. Some studies attributed high NO_X emissions of biodiesel to a single influencing factor. This increased NO_X tendency was attributed to the increased oxygen levels of biofuels, which leads to a higher cylinder temperature [45–47].

The oxygen levels of biodiesel result in an increase of the production of nitrogen oxides. The results of the experimental measurements performed on a diesel engine done by [48] probed the relationship between nitrogen oxide levels and oxygen content in fuel.

5. Conclusions

Currently, when our environment is overburdened with emissions of all kinds, the idea of using fuels with minimal impact on the environment is becoming very important. Fuels made from the methyl ester of seed oil can be considered advantageous mainly from the point of view that almost every diesel engine is, in principle, capable of burning such fuels. If we take into account the fact that up to 90% of the transport of people and goods is currently carried out by means of diesel vehicles (trucks, buses, locomotives, ships, tractors), it represents a huge potential. In addition, there are a large number of passenger cars equipped with diesel engines, which could also use fuels made from the methyl ester of seed oils for propulsion. In the countries of the European Union, their number ranges from 15 to 40%. Very often, the contribution of fuels produced from plant origin in terms of creating a balance of carbon dioxide circulation in nature is emphasized. The production of carbon dioxide during combustion corresponds to its consumption during photosynthesis. Biodegradability, e.g., rapeseed oil methyl ester after its release into the environment, is approximately 95% in 6 days.

The second part presents the results of the application of alternative fuels SAMPLE 1, SAMPLE 2, and their impact on the technical parameters, environmental parameters, and economy of the diesel engine. Based on the results of the measurement, their evaluation, as well as other significant facts that were found, it can be stated that when using biofuels as an energy source for the propulsion of vehicles equipped with a diesel engine, there is a decrease in power in the range from 1.5 to 3.8%. As for the specific fuel consumption, it increased by a maximum of 12% with the tested SAMPLE 1 fuel and for SAMPLE 2, a maximum increase of 4% was recorded.

Other important indicators from an ecological point of view are smoke values as well as NO_X emission values. The measurement of smoke values was performed at steady state with the load of the diesel engine. When measuring smoke at steady state when the combustion engine is loaded, it was not possible to evaluate the measurements in regard to the small volume of the diesel engine and also the limited range of the analyzer on which the measurements were made. From an ecological point of view, the values of nitrogen oxide emissions are also important. NO_X emission values were recorded in ppm and subsequently converted to $g \cdot kW^{-1} \cdot h^{-1}$. Measurements of NO_X emissions were performed at steady state with the diesel engine loaded. Their increase was recorded, with the alternative fuel SAMPLE 1 by 4.5% and with the alternative fuel SAMPLE 2 by 7.42% compared to the diesel fuel. In this case, nitrogen oxides were recorded in ppm.

As has been mentioned several times, the exhaust emissions of a diesel engine have a negative impact on the environment as well as on humans. These are mainly emissions of pollutants, especially "greenhouse" gases, which cause the gradual irreversible warming of the planet and upset the balance in nature with acid rain. This is mainly carbon dioxide CO_2, nitrogen oxides NO_X, methane CH_4, and sulfur oxides SO_X. Promoting the use of alternative fuels in transport is a step towards greater use of biomass, which will allow for a wider improvement of alternative fuels in the future, without excluding other options and in particular the possibility of using hydrogen.

Author Contributions: Conceptualization, J.J.; writing—original draft preparation, J.J.; methodology, J.J., J.T. and Ľ.H.; formal analysis, Z.T., P.F. and J.K.; writing—review and editing, P.F. and J.T.; visualization, Ľ.H., P.K. and M.T. All authors have read and agreed to the published version of the manuscript.

Funding: This research received no external funding.

Institutional Review Board Statement: Not applicable.

Informed Consent Statement: Not applicable.

Acknowledgments: This work was supported by AgroBioTech Research Centre built in accordance with the project Building "AgroBioTech" Research Centre ITMS 26220220180. This publication was supported by the Operational Program Integrated Infrastructure within the project: Demand-driven research for the sustainable and innovative food, Drive4SIFood 313011V336, cofinanced by the European Regional Development Fund.

Conflicts of Interest: The authors declare that they have no known competing financial interests or personal relationships that could have appeared to influence the work reported in this paper.

References

1. Markulík, S.; Šolc, M.; Petrík, M.; Balážiková, M.; Blaško, P.; Kliment, J.; Bezák, M. Application of FTA Analysis for Calculation of the Probability of the Failure of the Pressure Leaching Process. *Appl. Sci.* **2021**, *11*, 6731. [CrossRef]
2. Markulík, S.; Nagyová, A.; Turisová, R.; Vilinský, T. Improving Quality in the Process of Hot Rolling of Steel Sheets. *Appl. Sci.* **2021**, *11*, 5451. [CrossRef]
3. Official Statistics of Finland (OSF). *Greenhouse Gases*; Official Statistics of Finland (OSF): Helsinki, Finland, 2020; ISSN 1797-6065.
4. Tkáč, Z.; Graduš, J.; Jablonický, J.; Abrahám, R.; Bohát, M. *Alternatívne Palivá pre Motory*; Slovak University of Agriculture in Nitra: Nitra, Slovakia, 2008; ISBN 978-80-552-0095-8.
5. Jablonický, J.; Tkáč, Z.; Tulík, J.; Uhriniva, D.; Polerecký, J. *Automobilové Spaľovacie Motory*; Slovak University of Agriculture in Nitra: Nitra, Slovakia, 2019; ISBN 978-80-552-2015-4.
6. Qi, D.H.; Chen, H.; Geng, L.M.; Bian, Y.Z.H. Experimental studies on the combustion characteristics and performance of a direct injection engine fueled with biodiesel/diesel blends. *Energy Convers. Manag.* **2010**, *51*, 2985–2992. [CrossRef]
7. Tat, E.M.; Gerpen Van, J.H.; Wang, P.S. Fuel Property Effects on Injection Timing, Ignition Timing, and Oxides of Nitrogen Emissions from Biodiesel-Fueled Engines. Agris: International Information System for the Agricultural Science and Technology. 2011. Available online: http://web.cals.uidaho.edu/biodiesel/files/2013/08/Trans-50_4_1123.pdf (accessed on 21 September 2021).
8. Armas, O.; Yehliu, K.; Boehman, A.L. Effect of alternative fuels on exhaust emissions during diesel engine operation with matched combustion phasing. *Fuel* **2010**, *89*, 438–456. [CrossRef]
9. *ACEA Biodiesel Guidelines*; European Automobile Manufacturers Association: Brussels, Belgium, 2009.
10. Gums, M.; Kasifoglu, S. Performance and emission evaluation of compression ignition engine using a biodiesel (apricot seed kernel oil methyl ester) and its blends with diesel fuel. *Biomass Energy* **2010**, *34*, 134–139. [CrossRef]
11. Lenď'ák, P.; Švec, J.; Jablonický, J. Application of measurement methods of solid particle emission of diesel engine and their mutual comparation. *Acta Technol. Agric.* **2007**, *2*, 29.
12. Králik, M.; Jablonický, J.; Nikolov, M.I. *Monitoring of NO$_X$ Emissions at Selected Diesel Engine*; University of Ruse "Angel Kanchev": Ruse, Bulgaria, 2015; ISBN 978-954-712-678-7.
13. Šmerda, T.; Čupera, J.; Fajman, M. *Vznětové Motory Vozidel. Biopaliva, Emise, Traktory*; CPress Brno: Brno, Czech Republic, 2013; ISBN 978-80-264-0160-5.
14. Tauzia, X.; Maiboom, A.; Shah, S.R. Experimental study of inlet manifold water injection on combustion and emissions of an automotive direct injection diesel engine. *Energy* **2010**, *35*, 3628–3639. [CrossRef]
15. Hammond, G.P.; Kallu, S.; McManus, M.C. Development of biofuels for the UK automotive market. *Appl. Energy* **2007**, *85*, 506–515. [CrossRef]
16. Utlu, Z.; Kocak, M.S. The effect of biodiesel fuel obtained from waste frying oil on direct injection diesel engine performance and exhaust emissions. *Renew. Energy* **2008**, *33*, 1936–1941. [CrossRef]
17. Dorado, M.P.; Ballesteros, E.; Arnal, J.M.; Gómez, J.; López, F.J. Exhaust emissions from a diesel engine fueled with transesterified waste olive oil. *Fuel* **2003**, *82*, 1311–1315. [CrossRef]
18. Gunasekar, P.; Manigandan, S.; TR, P.K. Hydrogen as the futuristic fuel for the aviation and aerospace industry—Review. *Aircr. Eng. Aerosp. Technol.* **2020**, *93*, 410–416. [CrossRef]
19. Tesfa, B.; Mishra, R.; Zhang, C.; Gu, F.; Ball, A.D. Combustion and performance characteristics of CI (compression ignition) engine running with biodiesel. *Energy* **2013**, *51*, 101–115. [CrossRef]
20. Mofijur, M.; Masjuki, H.H.; Kalam, M.A.; Atabani, A.E.; Shahabuddin, M.; Palash, S.M.; Hazrat, M.A. Effect of biodiesel from various feedstocks on combustion characteristics engine durability and materials compatibility: A review. *Renew. Sustain. Energy Rev.* **2013**, *28*, 441–455. [CrossRef]
21. Abdulaziz, A.K.; Asad, S.; Abdallah, M.E.; Darshan, D.D.; Rajasree, S.; Kathirvel, B. Experimental assessment of performance, combustion and emission characteristics of diesel engine fueled by combined non-edible blends with nanoparticles. *Fuel* **2021**, *295*, 120590.
22. Manigandan, S.; Gunasekar, P.; Devipriya, J.; Nithya, S. Emission and injection characteristics of corn biodiesel blends in diesel engine. *Fuel* **2019**, *235*, 723–735. [CrossRef]
23. Balat, M.; Balat, H. Progress in biodiesel processing. *Appl. Energy* **2010**, *87*, 1815–1835. [CrossRef]
24. Dhas, A.A.; Casmir, A.; Sattar, S.; Amjad, S.M.; Joy, N.; Alagu, K.; Renita, A. Effect of mixing two biodiesels on emissions in CI engine fuelled by candle nut and soap nut methyl esters-diesel blends. *AIP Conf. Proc.* **2020**, *2311*, 020013.
25. Sharma, P.K.; Sharma, D.; Soni, S.L.; Jhalani, A.; Singh, D.; Sharma, S. Energy, exergy, and emission analysis of a hydroxyl fueled compression ignition engine under dual fuel mode. *Fuel* **2020**, *265*, 116923. [CrossRef]
26. Sarıkoç, S.; Örs, İ.; Ünalan, S. An experimental study on energy-exergy analysis and sustainability index in a diesel engine with direct injection diesel-biodiesel-butanol fuel blends. *Fuel* **2020**, *268*, 117321. [CrossRef]
27. Gad, M.S.; EL-Seesy, A.I.; Radwan, A.; He, Z. Enhancing the combustion and emission parameters of a diesel engine fueled by waste cooking oil biodiesel and gasoline additives. *Fuel* **2020**, *269*, 117466. [CrossRef]
28. Ramadhas, A.S.; Muraleedharan, C.; Jayaraj, S. Performance and emission evaluation of a diesel engine fueled with methyl esters of rubber seed oil. *Renew. Energy* **2005**, *30*, 1789–1800. [CrossRef]
29. Qi, D.H.; Geng, L.M.; Chen, H.; Bian, Y.Z.; Liu, J.J.; Ren, X.C. Combustion and performance evaluation of a diesel engine fueled with biodiesel produced from soybean crude oil. *Renew. Energy* **2009**, *34*, 2706–2713. [CrossRef]

30. Frijters, P.J.M.; Baert, R.S.G. Oxygenated fuels for clean heavy-duty engines. *Int. J. Veh. Des.* **2006**, *41*, 242–255. [CrossRef]
31. Jablonický, J.; Hujo, Ľ. Laboratórne Testovacie Zariadenia na Sledovanie Prietokovej Účinnosti Filtrov pri Aplikácii Biopalív so Simulovaním Prevádzkových Podmienok. Patent SK 288629 B6, 8 January 2019.
32. Ranjeet, K.R.; Sahoo, R.R. Engine performance, emission, and sustainability analysis with diesel fuel-based Shorea robusta methyl ester biodiesel blends. *Fuel* **2021**, *292*, 120234.
33. Dhar, A.; Kevin, R.; Agarwal, A.K. Production of biodiesel from high-FFA neem oil and its performance, emission and combustion characterization in a single cylinder DICI engine. *Fuel Process. Technol.* **2012**, *97*, 118–129. [CrossRef]
34. Can, Ö. Combustion characteristics, performance, and exhaust emissions of a diesel engine fuelled with a waste cooking oil biodiesel mixture. *Energy Convers. Manag.* **2014**, *87*, 676–686. [CrossRef]
35. Jablonický, J.; Tkáč, Z.; Majdan, R.; Uhrinová, D.; Hujo, Ľ.; Vozárová, V. *Properties Evaluation of Biofuels and Bio-Lubricants*; Slovak University of Agriculture in Nitra: Nitra, Slovakia, 2012; ISBN 978-80-552-0766-7.
36. Rataj, V. *Preparing a Scientific and Professional Text*; Slovak University of Agriculture in Nitra: Nitra, Slovakia, 2003; ISBN 80-8069-162-2.
37. Palenčár, R.; Wimmer, G.; Palenčár, J.; Witkovský, V. *Navrhovanie a Vyhodnocovanie Meraní*; Slovak University of Technology: Bratislava, Slovakia, 2021; ISBN 978-80-227-5080-6.
38. Aydin, H.; Bayindir, H. Performance and Emission Analysis of Cottonseed Oil Methyl Ester in a Diesel Engine. *Renew. Energy* **2010**, *35*, 588–592. [CrossRef]
39. Al-Widyan, M.I.; Tashtoush, G.; Abu-Qudais, M. Utilization of ethyl ester of waste vegetable oils as fuel in diesel engines. *Fuel Process. Technol.* **2002**, *76*, 503–521. [CrossRef]
40. Kalligeros, S.; Zannikos, F.; Stournas, S.; Lois, E.; Anastopoulos, G.; Teas, C.; Sakellaropoulos, F. An investigation of using biodiesel/marine diesel blends on the performance of a stationary diesel engine. *Biomass Bioenergy* **2003**, *24*, 141–149. [CrossRef]
41. Özgünay, H.; Çolak, S.; Zengin, G.; Sari, Ö.; Sarikahya, H.; Yüceer, L. Performance and emission study of biodiesel from leather industry pre-fleshings. *Waste Manag.* **2007**, *27*, 1897–1901. [CrossRef]
42. Kwanchareaon, P.; Leungnaruemitchai, A.; Jai-In, S. Solubility of a diesel–biodiesel–ethanol blend, its fuel properties, and its emission characteristics from diesel engine. *Fuel* **2007**, *86*, 1053–1061. [CrossRef]
43. Lapuerta, M.; Armas, O.; Rodríguez, F.J. Effect of biodiesel fuels on diesel engine emissions. *Prog. Energy Combust. Sci.* **2008**, *4*, 198–223. [CrossRef]
44. Wu, F.; Wang, J.; Chen, W.; Shuai, S. A study on emission performance of a diesel engine fueled with five typical methyl ester biodiesels. *Atmos. Environ.* **2009**, *43*, 1481–1485. [CrossRef]
45. Tamilselvan, P.; Nallusamy, N.; Rajkumar, S. A comprehensive review on perfor mance, combustion and emission characteristics of biodiesel fuelled diesel engines. *Renew. Sustain. Energy Rev.* **2017**, *79*, 1134–1159. [CrossRef]
46. Patel, R.L.; Sankhavara, C.D. Biodiesel production from Karanja oil and its use in diesel engine: A review. *Renew. Sustain. Energy Rev.* **2017**, *71*, 464–474. [CrossRef]
47. Jiaqiang, E.; Pham, M.; Zhao, D.; Deng, Y.W.; Le, D.H.; Zuo, W.; Zhu, H.; Liu, T.; Peng, Q.G.; Zhang, Z.Q. Effect of different technologies on combustion and emissions of the diesel engine fueled with biodiesel: A review. *Renew. Sustain. Energy Rev.* **2017**, *80*, 620–647.
48. Labeckas, G.; Slavinskas, S. The effect of rapeseed oil methyl ester on directinjection diesel engine performance and exhaust emissions. *Energy Convers. Manag.* **2006**, *47*, 1954–1967. [CrossRef]

Journal of
Marine Science and Engineering

Article

Influence of the Shape of Gear Wheel Bodies in Marine Engines on the Gearing Deformation and Meshing Stiffness

Silvia Maláková *, Michal Puškár, Peter Frankovský, Samuel Sivák and Daniela Harachová

Faculty of Mechanical Engineering, Technical University of Košice, Letná 9, 042 00 Košice, Slovakia; michal.puskar@tuke.sk (M.P.); peter.frankovsky@tuke.sk (P.F.); samuel.sivak@tuke.sk (S.S.); daniela.harachova@tuke.sk (D.H.)
* Correspondence: silvia.malakova@tuke.sk; Tel.: +421-055-6022372

Abstract: The basic properties of gears must be considered: the shape of their gearing, their load capacity, and the meshing stiffness, which affects the noise and vibration. When designing large gears, it is important to choose the correct shape of the gear body. Large gears used in marine gearboxes must be designed with as little weight as possible. The requirements of sufficient stiffness of the gear wheel body, as well as the meshing stiffness, must be met. This paper is devoted to the influence of spur gear wheel body parameters on gearing deformation and meshing stiffness. The stiffness of the gear is solved on the basis of the deformation of the gearing teeth, which is determined by the finite element method. Examples of the simulation and subsequent processing of results demonstrates how the individual parameters of the gear wheel body influence the stiffness of the gearing teeth. At the same time, the results point to designs of suitable shape and dimensions to achieve the required stiffness of the gearing teeth, but with the lowest possible weight of the spur gear wheel body.

Keywords: marine gearboxes; gear wheel shapes; forged spur gear; casted spur gear; meshing stiffness; gear teeth deformation; FEM

Citation: Maláková, S.; Puškár, M.; Frankovský, P.; Sivák, S.; Harachová, D. Influence of the Shape of Gear Wheel Bodies in Marine Engines on the Gearing Deformation and Meshing Stiffness. *J. Mar. Sci. Eng.* **2021**, *9*, 1060. https://doi.org/10.3390/jmse9101060

Academic Editor: María Isabel Lamas Galdo

Received: 24 August 2021
Accepted: 21 September 2021
Published: 26 September 2021

1. Introduction

The development of modern machines and the means of production are characterized by ever-increasing performance parameters with a reduction in equipment weight. This way of designing transmission mechanisms is common in the shipbuilding industry [1]. Reducing the weight of the gears to a minimum leads to higher stress values in machine parts, causing the formation of various failures. During the operation of gears, malfunctions arise, which can be caused by the following: the bending fatigue of gear teeth, the formation and expansion of pittings, the wear of the gear teeth by abrasion, gear tooth seizure, and insufficient tooth stiffness. Another group of malfunctions are gear failures, which can be caused by the following: the insufficient quality of gear material, gear manufacturing technology, and gear heat treatment; incorrect gear manufacturing; improper lubrication and incorrect lubricant selection; a foreign body in the lubricant; the incorrect mounting of the gears, and the wrong design of the gears with regard to the transmitted power.

The gearbox is one of the most commonly used mechanical components, which is widely used in mechanical transmission equipment. Gearboxes are manufactured with weights ranging from 20 kg up to more than 20 tons. The weight of marine gearboxes is in the heavier section of this range and they are able to produce torque of up to 1,100,000 Nm. The range of applications as well as the range of torques is very broad. Unlike a vehicle gearbox, the vast majority of marine gearboxes do not switch between cogs because there is no clutch to disconnect the drive. The marine gears are always in mesh and usually always turning [2]. They are engaged by some form of clutch locking them to the shaft or a brake band stopping a drum containing gears from turning (so the gears are forced to turn).

During the design of the geometry of the gears one can prevent errors through a correct and thorough calculation of these gears. Modern developments bring more precise computational methods based on new theoretical and experimental research. These methods are expected to determine the dimensions of the gears so that the gears achieve a high operational reliability. These methods should prevent the gear teeth from breaking under a given load or deforming beyond the functional level.

Under a given load, the teeth of the co-engaging wheels deform and this causes the failure of the meshing. The deformation of a tooth is generally quantified by the amount of tooth stiffness, which is atidefined as the load to deformon ratio. When designing gear parameters, it is an important step to focus on achieving the desired tooth stiffness.

The transmission mechanism is an acoustically closed system from which noise is propagated mainly by vibrations of the box surface or connected units, including the base structure [3,4]. The most significant cause of noise is the so-called Transmission Error. This error is related to the kinematic accuracy and stiffness of the gear.

The vibrations from the gears transmitted to the gearbox housing are the most significant source of noise [5–7]. From a physical point of view, the cause of vibrations is a dynamic force that can change its amplitude, direction, or field of action. In the case of involute gearing, the most significant change is the amplitude, the main cause of which is the variable stiffness of gearing and shock at the entry of teeth into meshing due to deformations and deviations in the pitch and profile of the teeth, which are different from the theoretical pitch and profile.

Due to the complex shape of the teeth, the theoretical determination of the deformation and stiffness of the teeth is difficult to carry out. Nowadays, it is characterized by a rapidly evolving computational technique that for solving a wide performs extensive calculations, where various numerical methods to solve gear problems are encountered. One of these methods is the finite element method. It is one of the numerical methods of mathematics, widely used for solving problems of elasticity and strength, dynamics of flexible bodies and, at the same time, for investigating the deformation of gears, and thus the stiffness during meshing.

During the design of the large gears, which are used in marine machines, it is also necessary to consider the influence of the shape of the gear body. Such a body must meet the basic requirements of stiffness and strength with the lightest possible construction of the gear body. The shape and geometrical parameters of the gear bodies used in marine gearboxes depend on the method of manufacture of these large gears. Gear wheels for marine gearboxes can be manufactured as castings, forgings and sometimes as welded gear body structures.

Several studies are aimed at the suitable geometric optimization of the shape of the gear body. Wang, in his work aimed at volume reduction in gear wheels, using the Matlab simulation, achieved a total of 25.2% mass reduction and 11.42% power loss of the optimized transmission [8]. The optimization was carried out from solid structure gear bodies with a combination of web-type structures with cross-shaped-type structures; the biggest gear wheel was optimized to the H-shape gear wheel structure. Monkova et al. carried out a study on the change of natural frequencies in gear wheel reduction [9]. In this study four gear wheels were made: one had a solid body and the next three had a ribbed body with the same count of ribs but a different rib width. Modal analysis proved that natural frequencies increased more slowly at low natural modes than at high natural modes. Xu Jin et al.'s work was to design a ball mill gear wheel [10]. This wheel was designed to have a smaller radius than the original gear wheel. Yet, the body shape of the mill gear wheel was designed to be lightweight using a profiled web connected to an outer gearing ring with holes of thicker width for connection to the ball mill. Naveen et al. conducted a study aimed at the thin-walled structures of gear wheels [11]. For this experiment three gears were designed: one with a thick symmetrical web, one with a thin asymmetrical web, and one with a thin symmetrical web. The results of the study revealed with lowering the web and rim thickness, the strength of the structure decreases. Marunic performed an analysis of a thin asymmetric thin-rimmed gear [12]. The conclusion of this study was that the maximum stress was located at the rim and web joint. This stress was at a maximum for the thinnest rim regarding the web thickness. However, for the thicker rim, the maximum stress increased with a thinner web. Zhao et al. made a structural optimization of a double web herringbone gear [13]. This gear was optimized by adding a material to the existing web structures to tilt them slightly. The volume of this gear increased by 5.3%, but the maximal stress in the gear decreased by 11.9%.

This paper is dedicated to the analysis of the basic parameters for the body of the spur gear, and their influence on the deformation and stiffness of the gear in the mesh. The design issue for large spur gears used in marine gearboxes, for which the semi-finished product is a casting or forging, is described in this paper. The stiffness of the teeth is solved based on the deformation of the gearing, which is determined by the finite element method. Examples of simulations and the subsequent processing of results demonstrate the profound influence of the individual parameters of the gear body on the gearing stiffness. At the same time, the results point to suitable designs of shape and dimensions to achieve the required gearing stiffness, but with the lowest possible weight of the spur gears.

2. Materials and Methods

Gears have various design types. In accordance with the careful design of the gearing, it is necessary to address the issue of designing a suitable shape of the gear wheel body. In the case of gears used in marine gearboxes, which are of large dimensions, the gears are designed to be lightweight. In the case of the necessary choice of a lightened gear body, it is necessary to maintain the stiffness of the body itself, but also to remember and consider the necessary meshing stiffness.

This work is focused on large-dimension spur gears made with a relief. Such gears can be forged, cast, or welded. The shape of the gear body depends on several factors, such as the size of the wheel, the material, the method of manufacture, and the use of the gear.

2.1. Body Shape of Forged Spur Gears

Forged spur gears are mainly used for the production of semi-finished products for gears, which will later be developed into their final form. These semi-finished products are formed by operations such as open-die drop forging or impression-die drop forging. The gearing is forged with additional material wrapped around the tooth profile. Both ferrous and non-ferrous metals such as carbon steel, alloy steel, stainless steel, titanium, nickel, aluminum, and tool steel can be used to produce forged gears.

Figure 1 shows the shapes of forged spur gears. Figure 1a shows the shape of the gear body without relief holes. The use of relief holes reduces the weight of the wheel. If the gearing width is less than the wheel hub width, the shape shown in Figure 1c is used. The most commonly used shape of forged spur gears is shown in Figure 1b. Gears with a diameter $d \leq 150$ to 200 mm are made of a solid bar semi-finished product by forging. Drop forging is used for spur gears with a diameter $d = 160$ to 200 mm. The design of the shape of the forged spur gear, together with the prescription of the basic characteristic dimensions of the wheel, such as the dimensions of the rim, arms and hub, are shown in Table 1.

For analyzing the influence of the body shape of the forged spur gear on the deformation and engagement stiffness, a gear with the number of teeth $z = 71$ and the normalized modulus value $m_n = 2.5$ mm was chosen. The basic dimensions for the variants of the investigated spur gear bodies that were selected are given in Table 2. Variant A1 is designed according to the dimensions shown in Table 1 and Variant A2 is a full wheel shape without weight relief. A gear body with an increased web thickness was used in Variant A3 ($f = 0.7b$); with a rim of twice the thickness and a shape without relief holes in Variant A4; with a rim of three times the thickness and a shape without relief holes in Variant A5; and with a rim of half thickness (see Table 1) and a shape without relief holes in Variant A6.

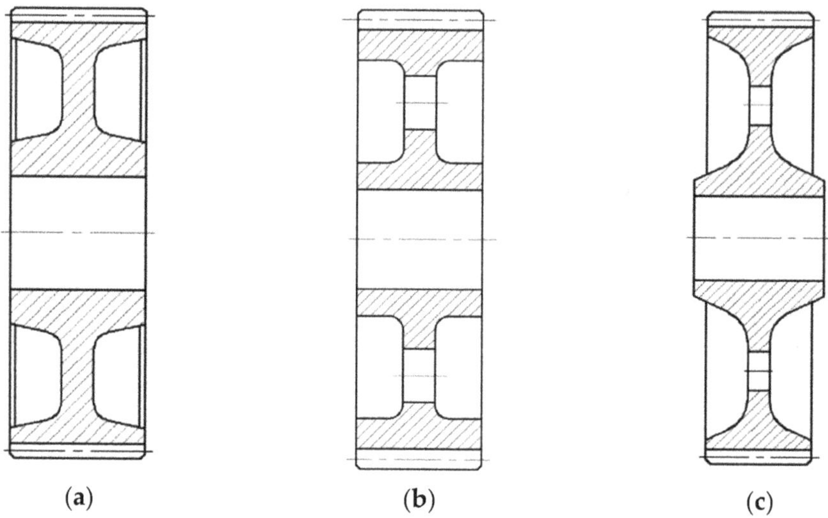

(a)   (b)   (c)

Figure 1. Body shapes of forged spur gears: (**a**) shape without relief holes; (**b**) the most commonly used shape of forged spur gears; (**c**) shape with openings for weight reduction.

Table 1. Forged spur gear—basic geometric parameters.

Forged Spur Gear	Gear Parameters
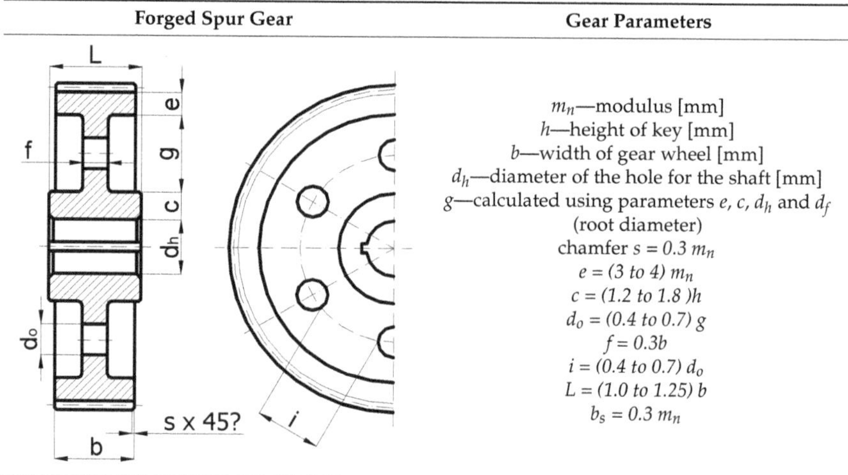	m_n—modulus [mm] h—height of key [mm] b—width of gear wheel [mm] d_h—diameter of the hole for the shaft [mm] g—calculated using parameters e, c, d_h and d_f (root diameter) chamfer $s = 0.3\, m_n$ $e = (3\ to\ 4)\, m_n$ $c = (1.2\ to\ 1.8\,)h$ $d_o = (0.4\ to\ 0.7)\, g$ $f = 0.3b$ $i = (0.4\ to\ 0.7)\, d_o$ $L = (1.0\ to\ 1.25)\, b$ $b_s = 0.3\, m_n$

Table 2. Variants of forged spur gears.

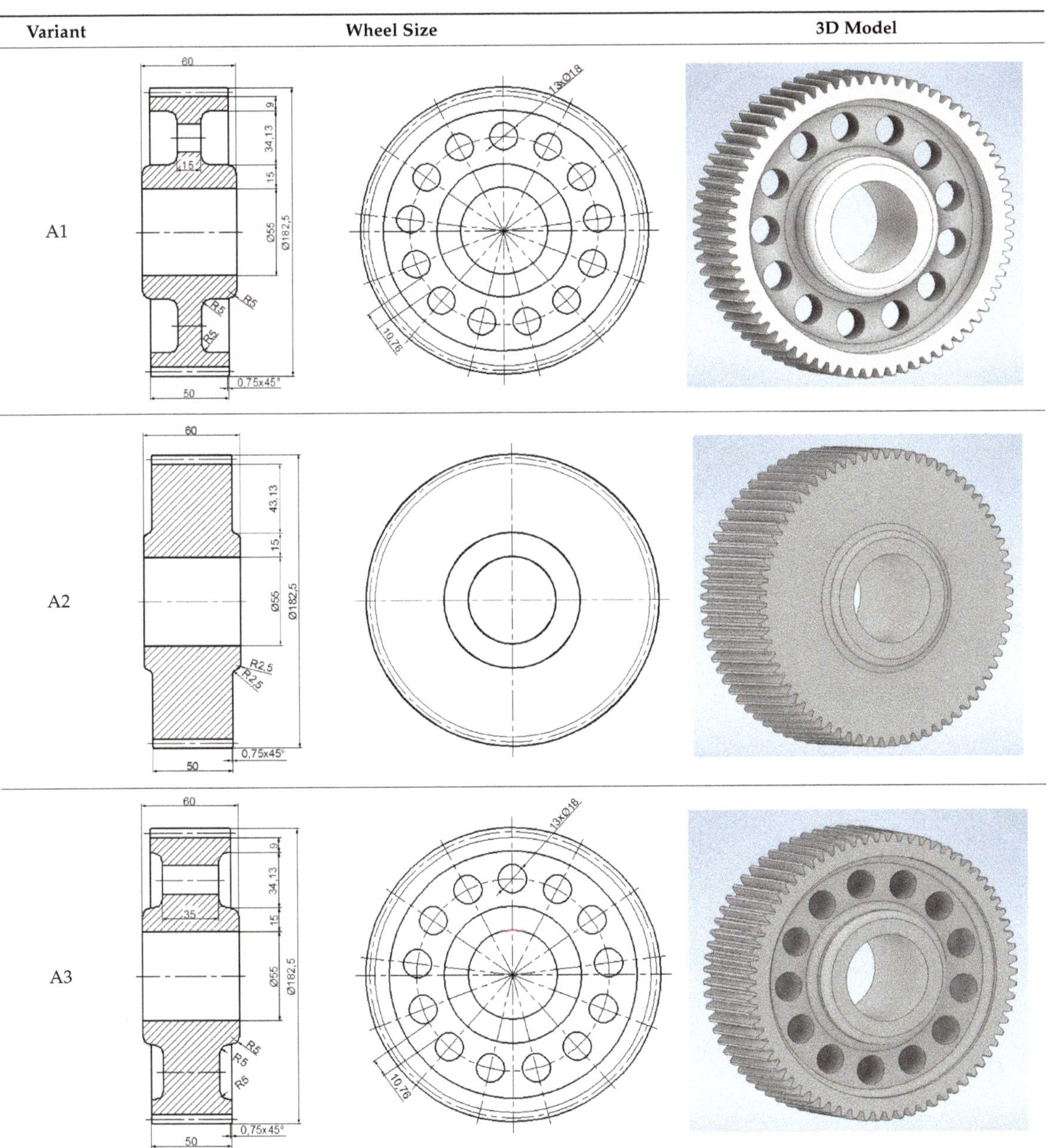

Variant	Wheel Size	3D Model
A1		
A2		
A3		

Table 2. *Cont.*

Variant	Wheel Size	3D Model
A4		
A5		
A6		

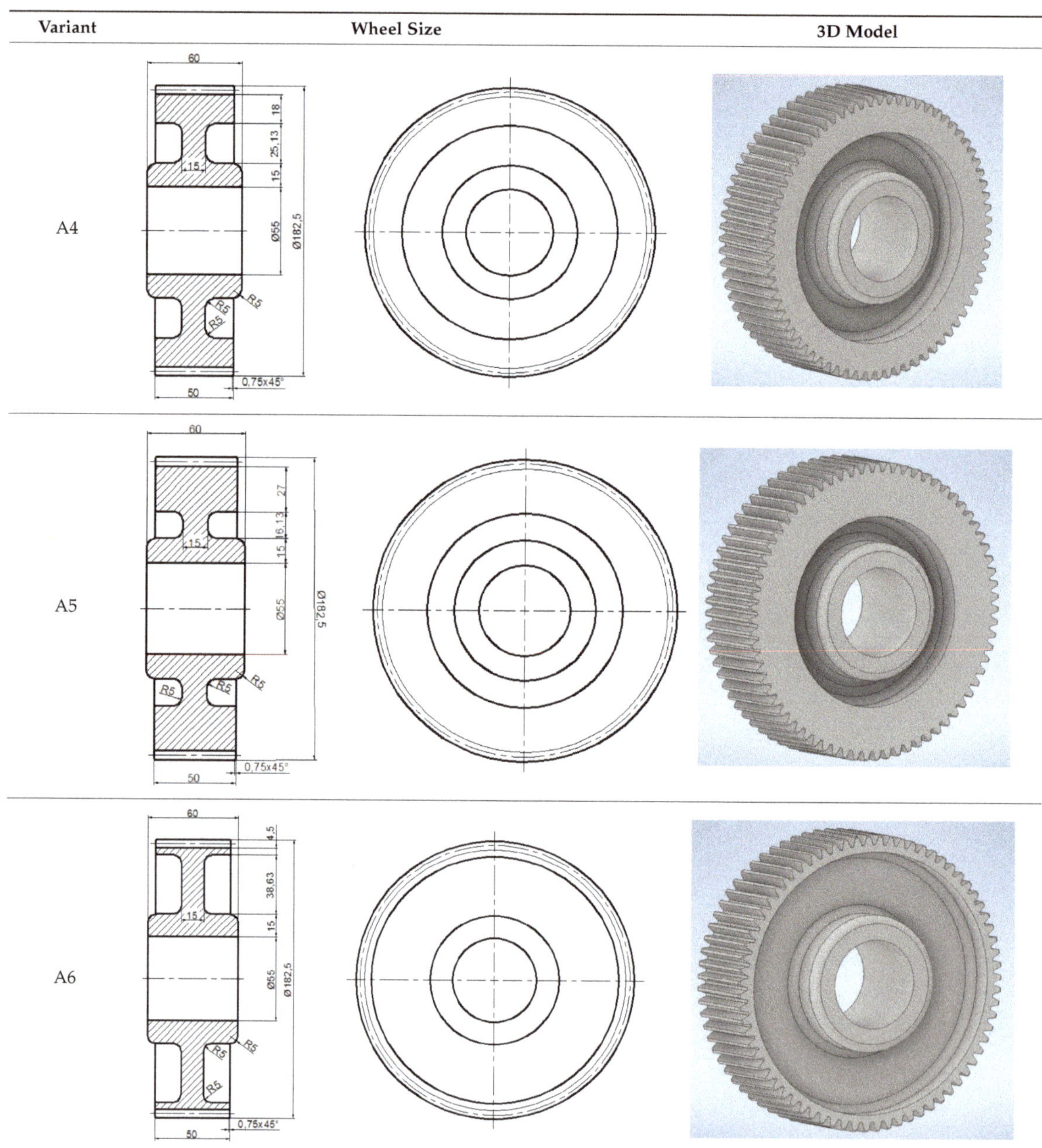

2.2. Construction of Cast Spur Gears

Casting is mostly used to create semi-finished products for large gear wheels. These gear wheels are less durable than forged wheels. The material used for these types of gear wheels is gray cast iron for a circumferential speed $v < 7$ m s^{-1} and cast steel for a circumferential speed $v < 20$ m s^{-1}. For spur gear diameter $d > 500$ mm, the shape given in Table 3 is used.

Table 3. Cast medium-sized spur gear: basic geometric parameters.

Cast Medium-Sized Gear Wheel	Wheel Parameters
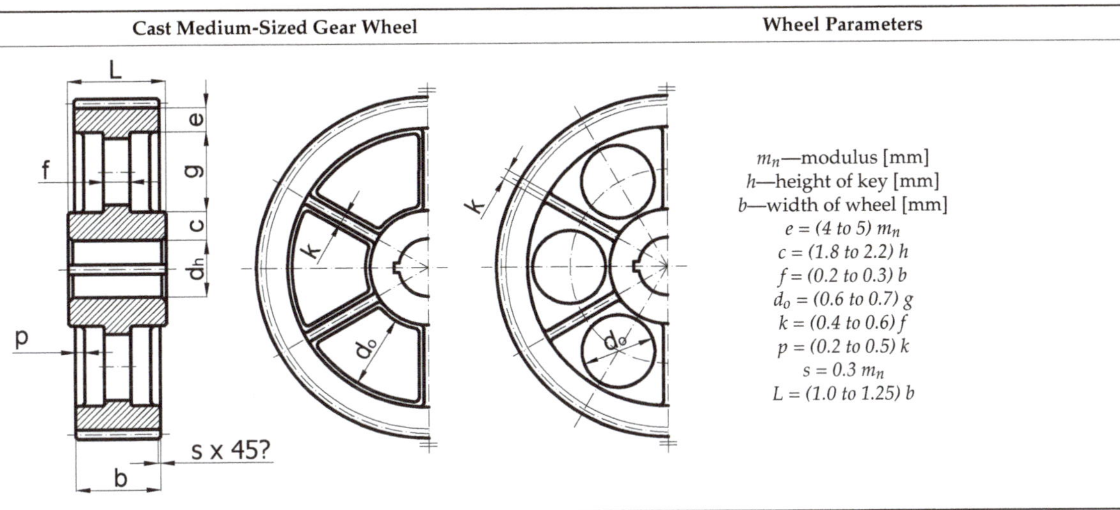	m_n—modulus [mm] h—height of key [mm] b—width of wheel [mm] $e = (4$ to $5) m_n$ $c = (1.8$ to $2.2) h$ $f = (0.2$ to $0.3) b$ $d_o = (0.6$ to $0.7) g$ $k = (0.4$ to $0.6) f$ $p = (0.2$ to $0.5) k$ $s = 0.3 m_n$ $L = (1.0$ to $1.25) b$

Cast wheels can also be made with a full disc. For very large cast wheels, the connection of the hub to the rim is realized by means of arms with an elliptical, cross-sectional T-section (see Table 4). The arms must be checked with calculations for bending strength.

Table 4. Large casted spur gear—basic geometric parameters.

Large Casted Spur Gear	Wheel Parameters
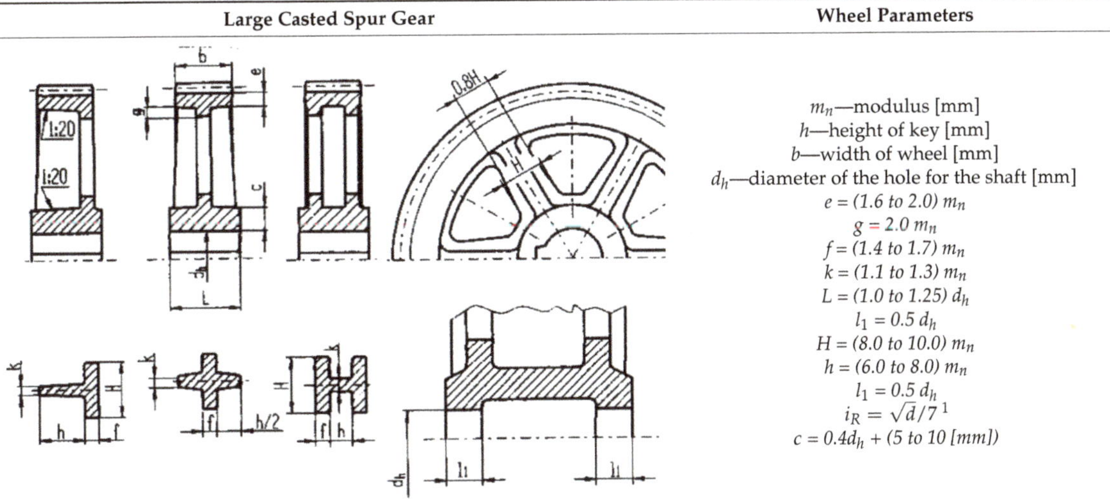	m_n—modulus [mm] h—height of key [mm] b—width of wheel [mm] d_h—diameter of the hole for the shaft [mm] $e = (1.6$ to $2.0) m_n$ $g = 2.0 m_n$ $f = (1.4$ to $1.7) m_n$ $k = (1.1$ to $1.3) m_n$ $L = (1.0$ to $1.25) d_h$ $l_1 = 0.5 d_h$ $H = (8.0$ to $10.0) m_n$ $h = (6.0$ to $8.0) m_n$ $l_1 = 0.5 d_h$ $i_R = \sqrt{d}/7$ [1] $c = 0.4 d_h + (5$ to 10 [mm]$)$

[1] d—diameter of the pitch circle [mm], i_R—number of arms, selected from 3 to 8.

To analyze the influence of the body shape of the cast spur gear on the deformation and meshing stiffness, a gearing with the number of teeth $z = 71$ and the standardized modulus value $m_n = 7$ mm, was chosen. The shapes of the investigated bodies of cast spur gears were designed, the basic dimensions for which are shown in Table 5. Variants of body shapes B1 and B2 are designed according to Table 3; a variants B3 and B4 are designed according to Table 5.

The work was created on a practical requirement basis to design the optimal shape of the body of the gear wheel with relief, i.e., gear wheels with the lowest possible weight, while maintaining the stiffness not only of the wheel body but also the stiffness and strength of the gear. The stiffness of the gearing is determined on the basis of the deformation of the gearing, whereas the deformation of the gearing is investigated by means of the finite element method.

Table 5. Variants of cast spur gears.

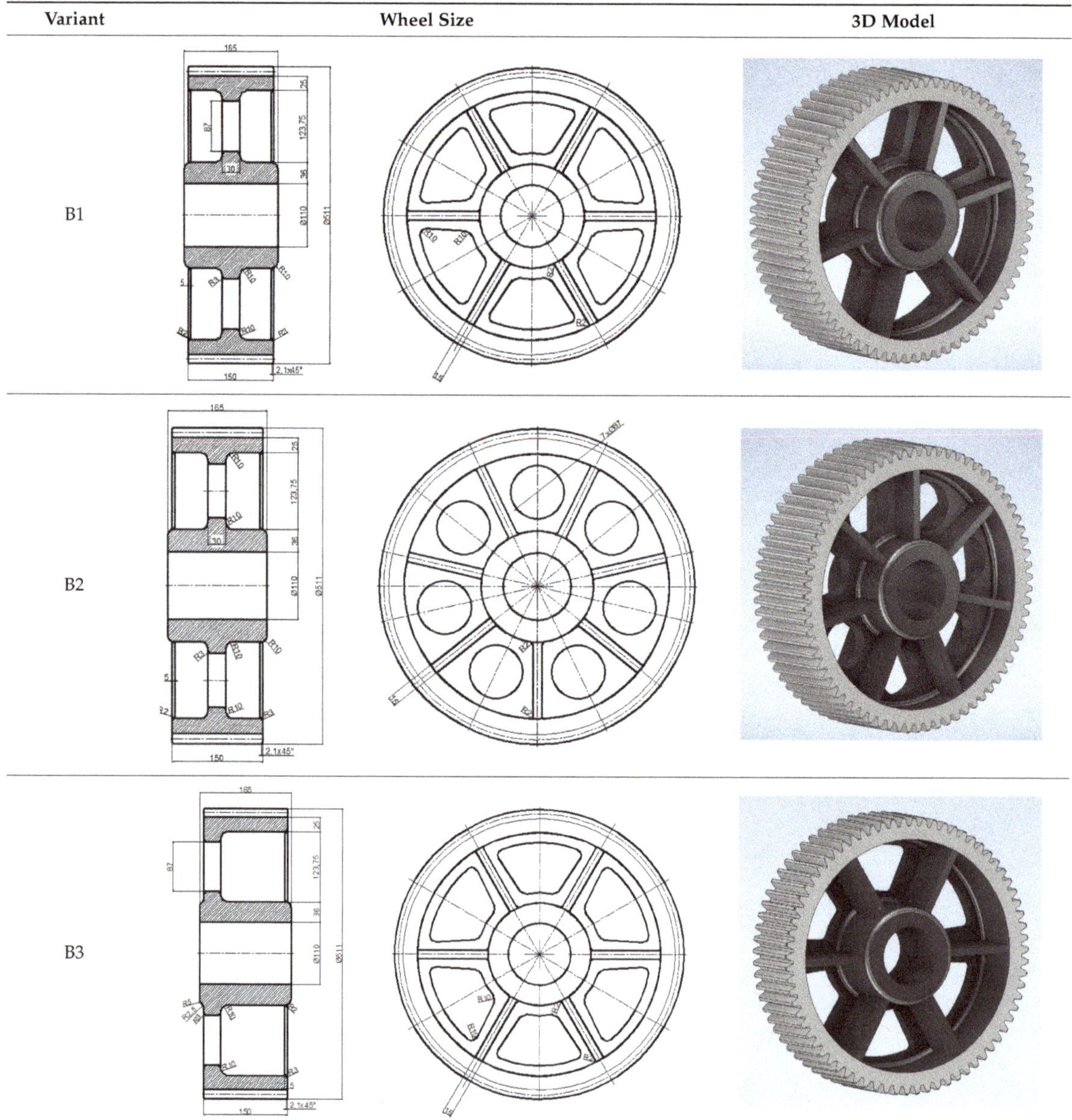

Variant	Wheel Size	3D Model
B1		
B2		
B3		

Table 5. *Cont.*

Variant	Wheel Size	3D Model
B4		

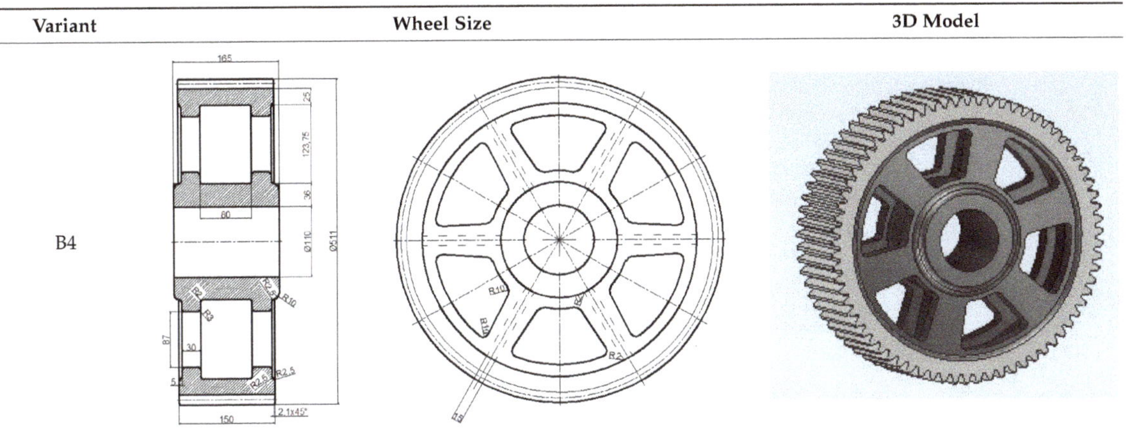

2.3. Gear Deformation and Meshing Stiffness

The teeth of the gears are deformed by the load. Under the action of the resulting normal force F, the tooth of one wheel is deformed, as illustrated by a thick line in Figure 2. The resulting deformation of the tooth in the direction of the normal force δ_i ($i = 1,2$—index, which distinguishes whether it is a pinion tooth, drive wheel or a driven wheel tooth) consists of a deformation from bending, shearing, the deformation at the point of entry and from tactile deformation [14].

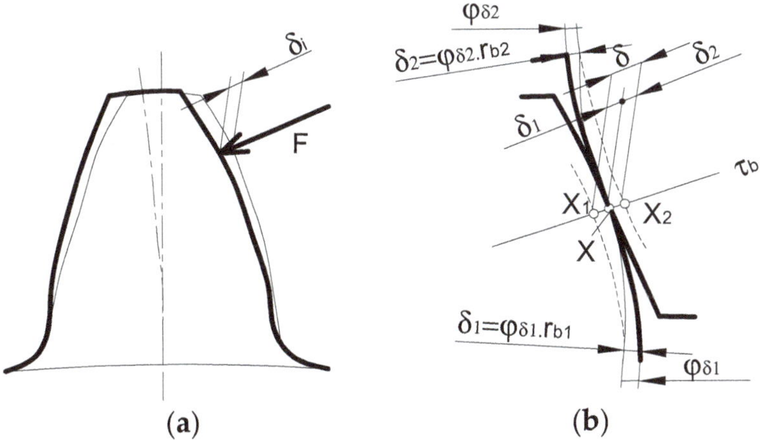

Figure 2. Deformation of (**a**) tooth, (**b**) teeth meshing deformation.

However, in reality, it is necessary to determine the deformation of the interlocking teeth, i.e., the deformation of a pair of teeth, which can be imagined and illustrated in Figure 2b. This figure shows a section of a pair of teeth, which touch at the point X on a line of action τ_b in the unloaded state. After loading, the profiles of the two teeth engaging together deformed into the shape shown by the dashed line in the respective figure. These already deformed tooth profiles intersected the line of action at points X_1 and X_2. The total deformation of this pair of teeth δ could then be determined as the sum of the deformations of both teeth δ_1 and δ_2. There are the angles $\varphi_{\delta 1}$ and $\varphi_{\delta 2}$ shown in Figure 2. These are the angles by which the individual wheels must turn to meet again at point X, as is in fact the case.

The deformation is the same as the stiffness of the individual pairs of teeth and varies along the line of action [15,16]. A tooth has the greatest deformation when a force acts on the tip of the tooth, due to the large deformation of the tooth caused by bending [17]. To solve the common tasks of gear strength calculation, shaft calculation, and gear bearing, as well as in the case of the gear stiffness solution, the continuous load (w_p) is replaced by isolated forces. The result of this continuous load is the force acting on the side of the tooth in the frontal plane. Because the gear ratio is greater than 1 for each gear, the resulting force acting on the tooth varies during the engagement depending on the number of tooth pairs engaging with each other. For precision spur gears with straight teeth, the load distribution between one and two pairs of teeth during the engagement is shown in Figure 3. In the two-pair engagement section (sections AB and DE), two pairs of teeth, I and II, are spaced apart by the basic pitch p_{tb}, and the load of individual pairs is directly proportional to the stiffness of these pairs with regard to common deformation [18,19]. In the single-pair engagement section BD, the load is transmitted by one pair of teeth, where ζ is the coordinate of the line of action.

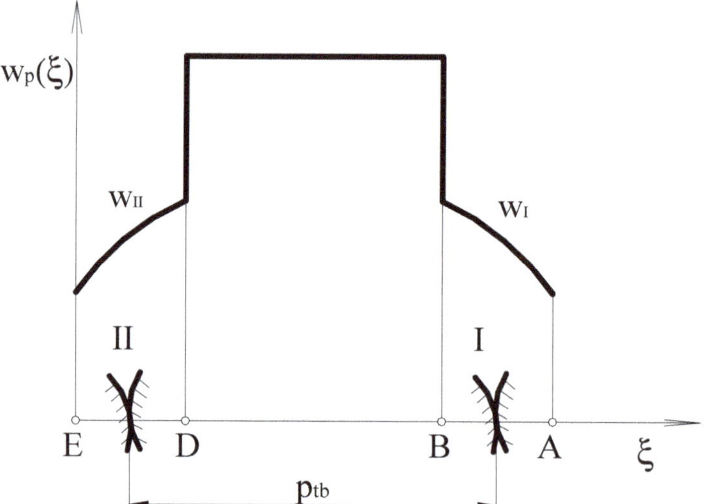

Figure 3. Load of individual pairs of teeth of precise straight spur gearing (sections AB and DE are two engaging gear pairs of teeth; section BD is one engaging gear pair of teeth).

For the study of mesh stiffness, the distribution of the load on individual pairs of teeth meshing together is very important [20–23]. The stiffness of the individual teeth is examined, not the meshing stiffness. The maximum tooth deformation occurs when a load acts on the addendum of the tooth, i.e., at point A (see Figure 3). In this section, the load is divided between two meshing pairs of teeth. The deformation of the teeth was examined to observe if the load acted on the tooth addendum and split it between two meshing pairs of teeth that were simultaneously engaged.

Deformations of teeth are generally quantified by tooth stiffness, which is defined as the ratio of the load to the deformation. We define tooth stiffness as the force per unit of length that is required for a deformation by 1 μm. During the engagement cycle, the stiffness of the gearing changes is mainly due to a change in the bending arm.

Gear stiffness results are determined based on gear deformation. Gear deformation is determined by using the finite element method. The condition for the successful mastering of this issue is the creation of a computational model for solving the problems of static deformation analysis by the finite element method. The first step is to create geometric 3D models of individual gear wheels. The deformation of the teeth can also be examined on a model of the entire gear wheel or only on a suitable part of the gear wheel. The deformation was investigated on the model of the whole gear wheel.

The next step was to define the material properties of the gear wheels. The material from which the gear wheel was made was replaced in the study for solving FEM tasks, by using material constants which characterize the material. Basic material properties such as Young's modulus of elasticity, the modulus of rigidity and Poisson's number were used.

Due to the properties and use of the elements, a finite element of the SOLID type was used for the shape of the investigated three-dimensional geometric model of the gear wheel. This element was the shape of a prism. Due to the shape of the model, as well as the accuracy of the results, an 8-node version of this element with curved edges was chosen. The density of the mesh was defined by entering the number of elements on the control curves of the mesh entities. Based on the research, the finer mesh on the areas of the examined teeth was used to obtain the required accuracy of the results to finding this solution. The finer mesh was created by entering an appropriate number of finite elements on the individual control curves, given the size of the model. Here, it was necessary to follow the principle of creating the same number of elements on the curves forming the boundaries of neighboring entities.

In the mechanics of flexible bodies, we recognized two types of boundary conditions, geometric and force. Geometric boundary conditions defined the displacements on a part of the boundary of the examined body. In the place of rigid, inflexible bonds, the nodal displacements or rotations were zero. These were defined on the gear wheel hole. The force boundary conditions represented the prescribed surface forces at the body boundary. The action of the co-engaging teeth was replaced by a continuous constant load.

To verify the results of tooth deformation solved by the finite element method, tooth deformation was experimentally determined by the static measurement of tooth deformation at the point of load with a constant force [24–32].

3. Results and Discussion

The analysis of how the shape of the spur gear body influences the gear meshing stiffness is solved based on the results of the tooth deformation. In this work, the deformation of the gear is solved using the finite element method. To accomplish this, a computational model of the examined gear is created, which is the basis for solving the problems of static deformation analysis by the finite element method, using the program SolidWorks.

The load on the teeth of the spur involute gear has a continuous character. To solve the common tasks of gear strength calculation, shaft calculation, and gear bearing, the continuous load is replaced by a single force. Maximum deformations were investigated when the load acted on the tooth addendum (Figure 4).

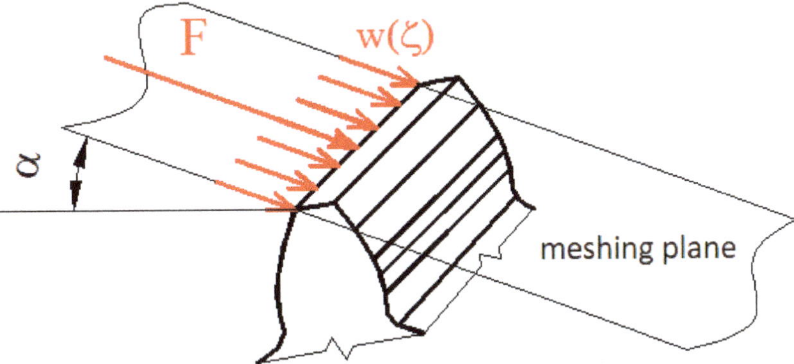

Figure 4. Expression of continuous load by single force.

3.1. The Deformation and Gear Stiffness of Forged Gear Wheels

The gearing on the forged wheels was with the number of teeth $z = 71$ and the normalized value of the modulus, $m_n = 2.5$ mm. The gear width was $b = 50$ mm and the problem was solved for spur gear wheels. The deformation was investigated for the wheel models listed in Table 2. The value of width load was $w = 100$ N/mm, expressed on the basis of a single force of $F = 5000$ N.

Figures 5 and 6 shows the distribution of deformation across the width of the tooth when a load is applied to the addendum of the tooth for individual models of forged gear wheel shapes. As can be seen from the figure, the distribution of the deformation along the gear width was comparatively the same for all the examined shapes of the forged gear bodies. An exception is the A6 model (see Table 2), whose rim thickness is the smallest. For the model A6, the thickness of the rim is equal to *1.8 m_n*, which is much less than the permissible minimum thickness of the rim (*3.5 m_n*). Therefore, the tooth deformation reflects the effect of a web with a thickness equal to $f = 0.3\,b = 15$ mm located in the middle of the width of the wheel. At the location of the web, the deformation of the tooth decreased due to the supporting material of the web. The tooth deformation for model A1 was higher than for models A2 to A5, because the thickness of the rim was slightly greater than the minimum rim thickness defined by the standard value.

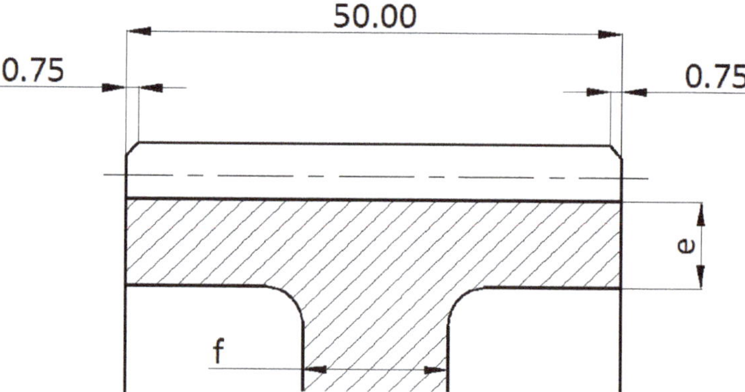

Figure 5. Active gear width during engagement of the wheels.

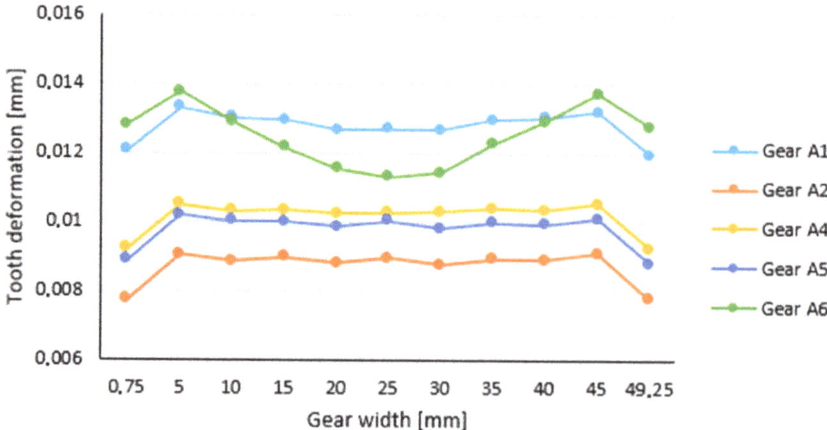

Figure 6. Comparison of tooth deformation along the width of the forged gear wheel for individual models.

The deformation distribution along the width of the gear is also applied in Table 6.

Table 6. Gear deformations at the load point for the designed models of forged wheels.

Variant	3D Model	Maximum Deformation [mm]	Illustration of Deformation	Cut Section
A1	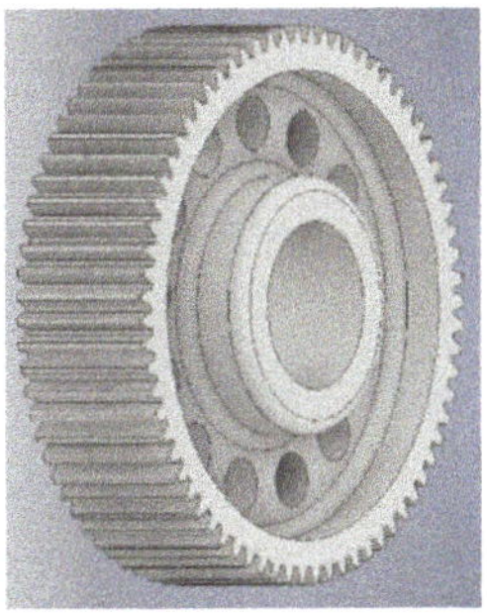	 0.01352 Max. 0.01082 0.00811 0.00541 0.0027 0 Min.	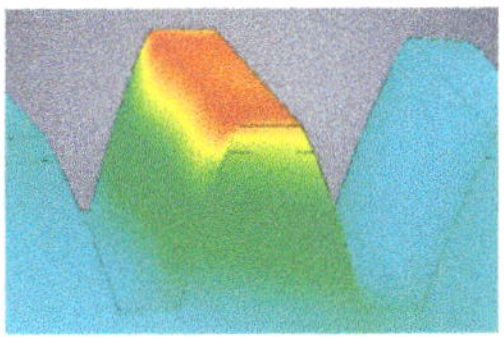	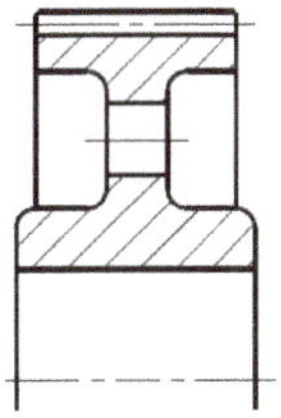
A2	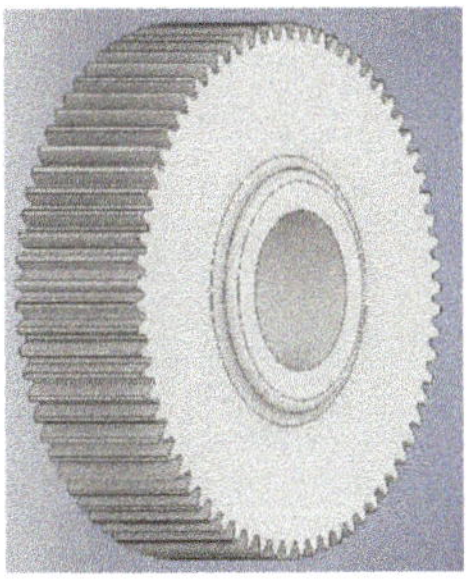	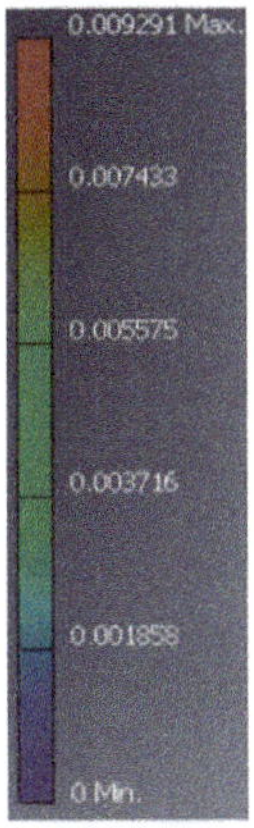 0.009291 Max. 0.007433 0.005575 0.003716 0.001858 0 Min.	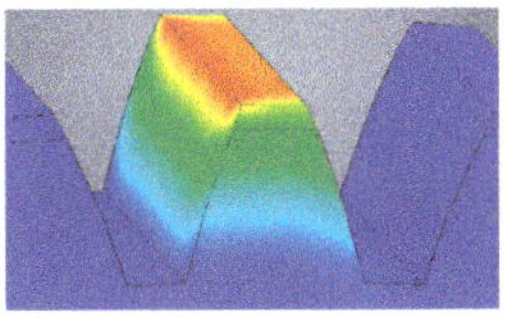	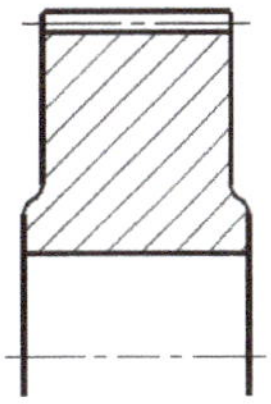
A3	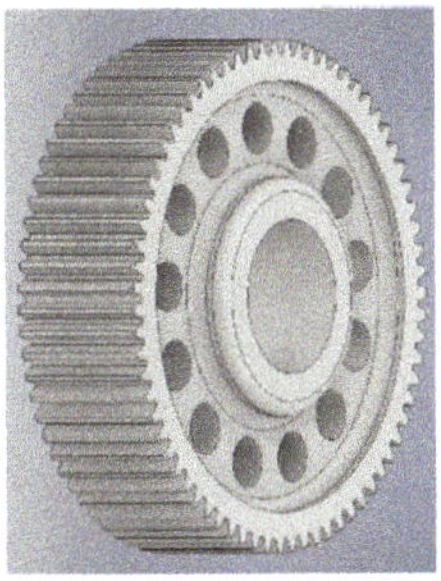	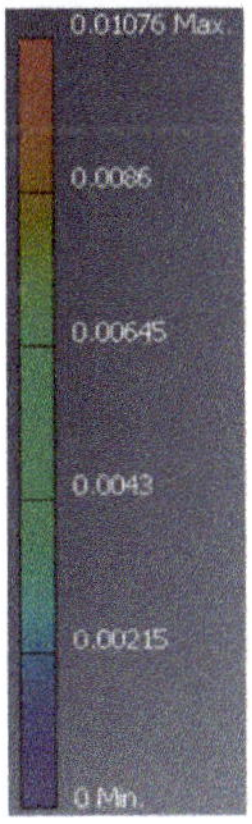 0.01076 Max. 0.0086 0.00645 0.0043 0.00215 0 Min.	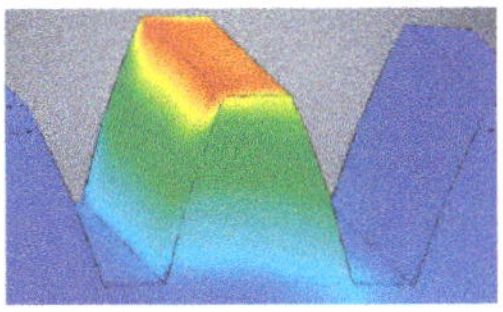	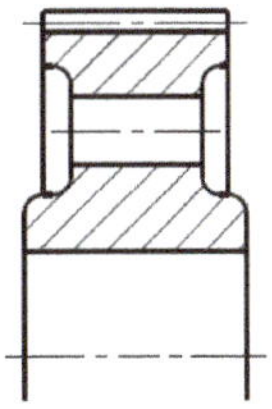

Table 6. *Cont.*

Variant	3D Model	Maximum Deformation [mm]	Illustration of Deformation	Cut Section
A4	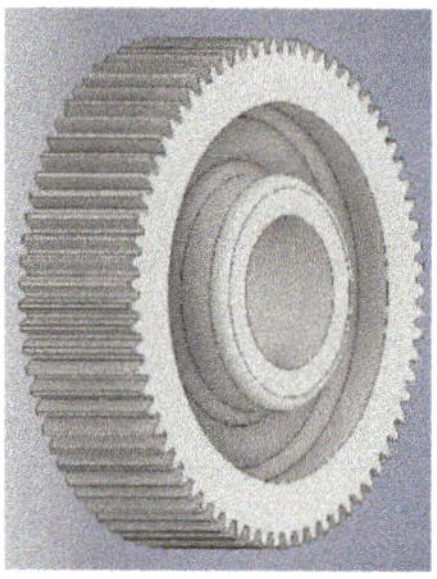	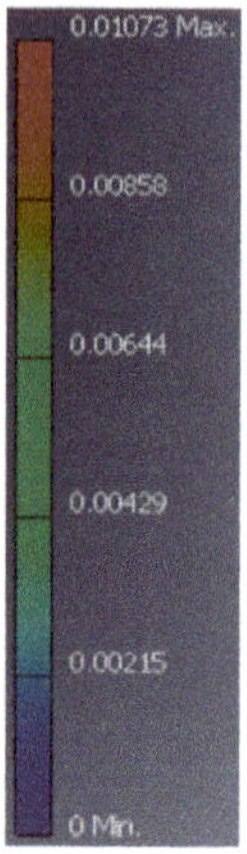 0.01073 Max. 0.00858 0.00644 0.00429 0.00215 0 Min.	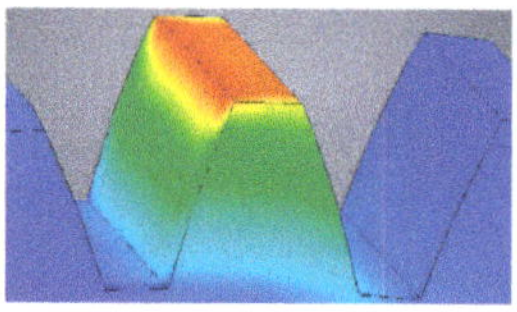	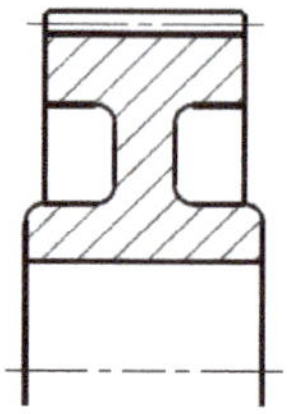
A5	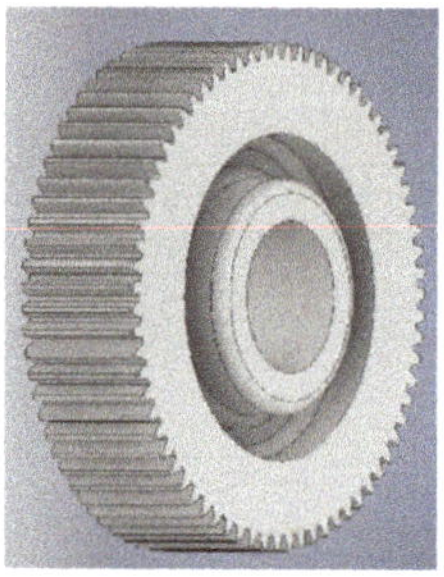	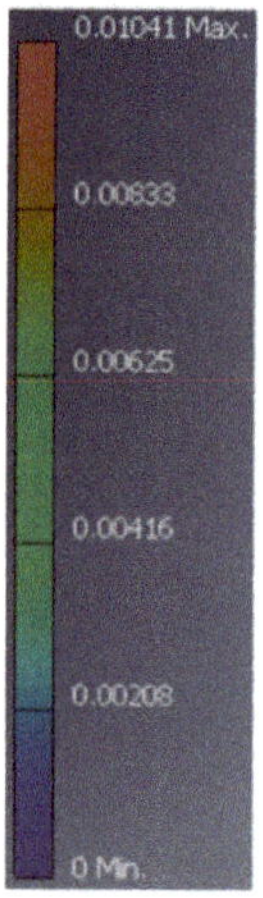 0.01041 Max. 0.00833 0.00625 0.00416 0.00208 0 Min.	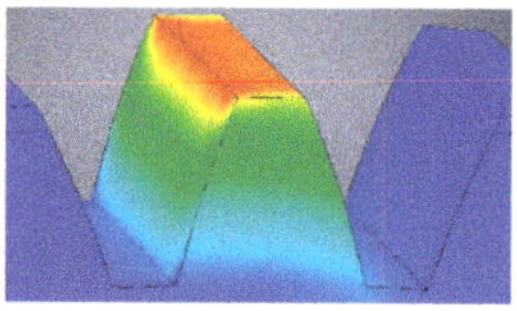	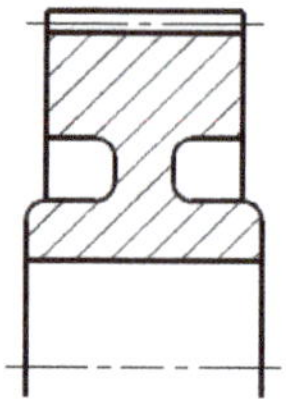
A6	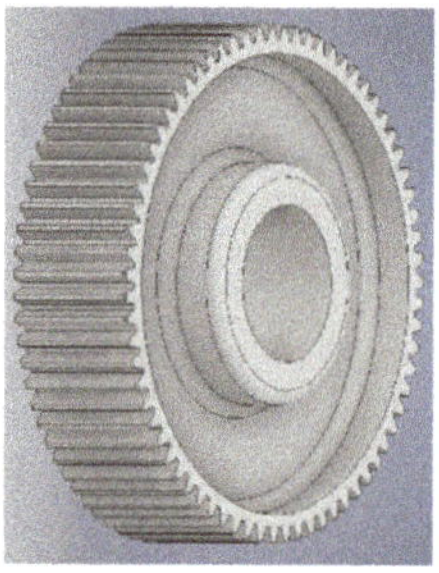	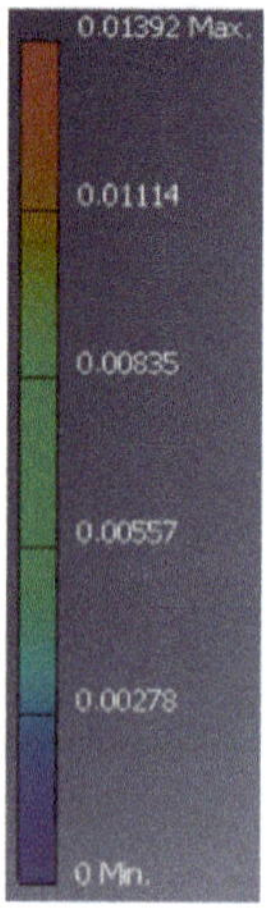 0.01392 Max. 0.01114 0.00835 0.00557 0.00278 0 Min.	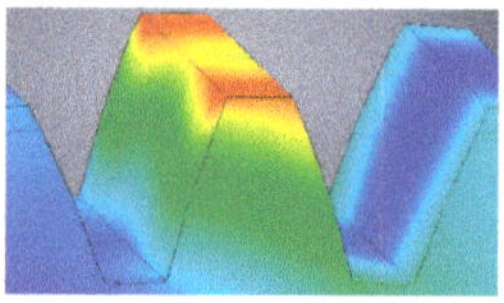	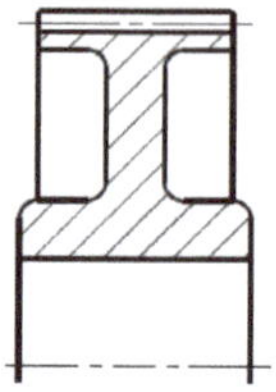

In the next part, the influence of the rim thickness on the tooth deformation was investigated. Several spur gear models were created, where the web thickness was the same for all models (value of $f = 15$ mm) and the rim thicknesses varied (see Figure 5). Since the deformation along the tooth width was not constant beyond the deformation value that represented the tooth deformation, the deformation at the center of the tooth width was defined. Figure 7 shows the dependence of the tooth deformation on the thickness of the rim. As the results show, reducing the thickness of the rim increased the deformation of the teeth.

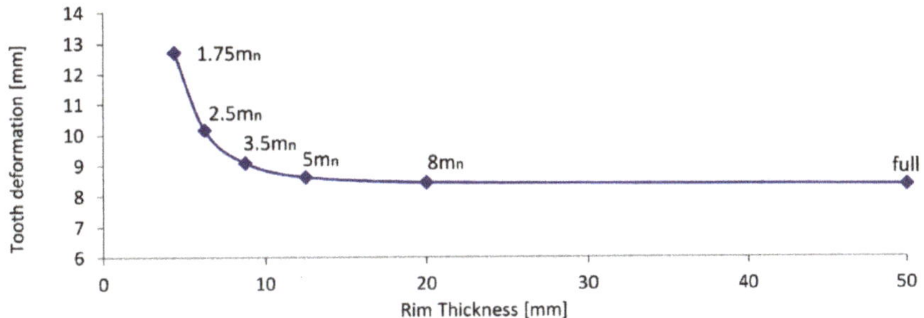

Figure 7. Influence of rim thickness on gear deformation.

As can be seen from the acquired values, up to a rim size of $3.5\,m_n$, the deformation changed by a large amount.

Figure 8 shows a comparison of the tooth stiffness over the width of the contact course for the individual shapes of the forged spur gears. The increase in stiffness (decrease in deformation) at the edges of the wheel was caused by a chamfer, i.e., a supporting effect. The gear wheel with a full body shape (variant A2) had the greatest stiffness. Variant A2 had the smallest deformation of the teeth, which was caused by the supporting effect of the material. Variant A5 was the model with the second smallest deformation of the teeth, and thus the species with the greatest stiffness of the teeth. The following are the A4 and A3 gears. There was very little difference between the values of tooth deformation (tooth stiffness) despite the different shapes of the bodies. The biggest difference in the stiffness distribution across the width had the smallest rim thickness for the forged body model (Variant A6).

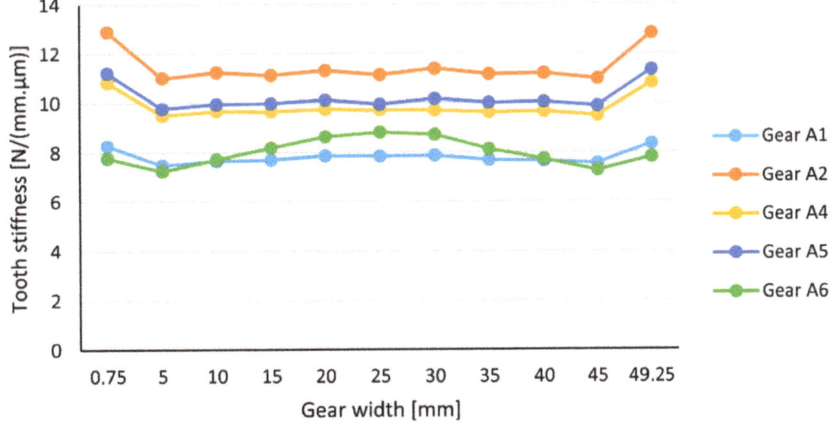

Figure 8. Comparison of gear stiffness across width for forged spur gears.

Figure 9 shows the effect of the rim width on gear stiffness. As the thickness of the rim increased, the rigidity of the gearing also increased. The thickness of the rim was of a lesser value than the *3.5 m_n* and affected the stiffness of the gearing more significantly; therefore, it was appropriate to choose the thickness of the rim equal to or greater than this value. For the A6 model, the thickness of the rim was equal to *1.8 m_n*, which was much lower than the permissible minimum thickness of the rim (*3.5 m_n*). Therefore, the stiffness of the tooth reflected the effect of a web of size *f = 0.3 m_n = 15* mm located in the middle of the wheel width. At the location of the web, the stiffness of the tooth increased due to the supporting material of the web, unlike other models, where the thickness of the rim is greater than the allowed minimum.

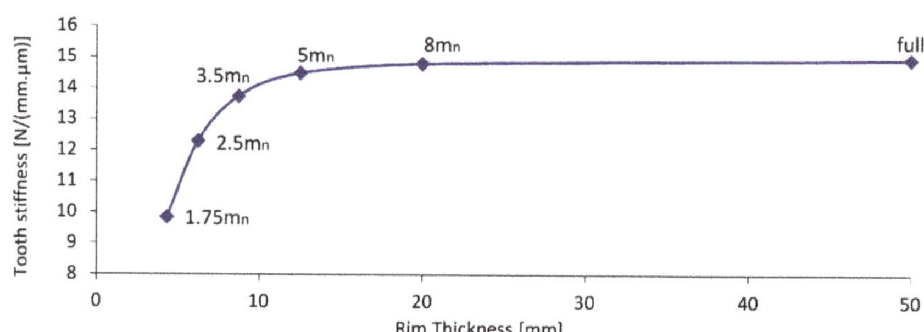

Figure 9. Influence of rim thickness on gear stiffness.

A comparison of the effect of material reduction on gear deformation is shown in Figure 10. The greater the mass reduction, then the greater the final gear deformation. By comparing the effect of the percentage reduction in mass on the deformation of the gearing, the forged wheels marked A3 and A4 appear to be the most suitable variants (Table 2). These wheels have very similar gear deformations and mass reductions, while they differ in shape from each other. The A4 variant appears to be more advantageous from the point of view of the economy than the A3 variant, which has a more complicated shape for production and requires an additional machining process by drilling relief holes.

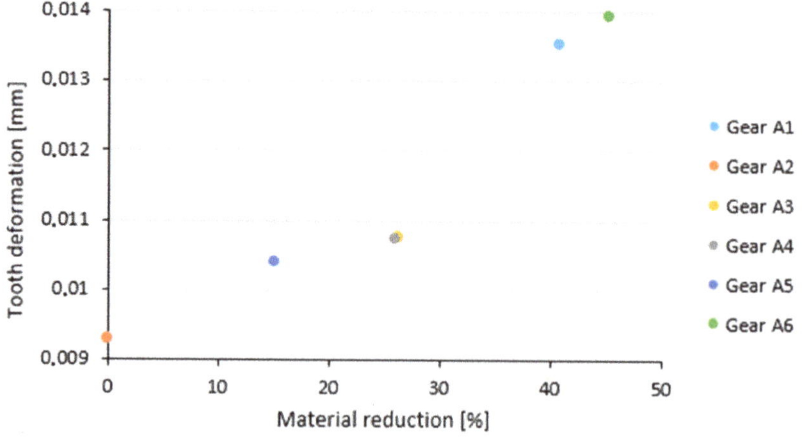

Figure 10. Influence of material reduction on maximum gear deformation.

3.2. Deformation and Stiffness of Gearing of Cast Gears

The gearing of the cast gear wheels had the number of teeth *z = 71*, the normalized value of the modulus *m_n = 7* mm, and a gear width of b = 150 mm. The problem was solved for spur gear wheels. The deformation was investigated for the wheel models listed in Table 5. The magnitude of the width load, w = 100 N/mm, was expressed on the basis of a single force, F = 15,000 N.

The first step was to investigate the distribution of deformation along the wheel width where it was found that the deformation of the gearing was unequal; the teeth of the wheel that were located above the rib had less deformation than the teeth that were located above the holes (see Figures 11–14). The smallest difference was found on model B2.

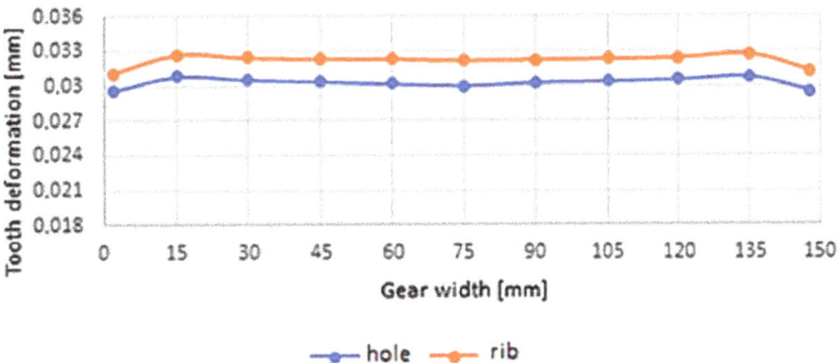

Figure 11. Deformation along the width of the tooth for model B1.

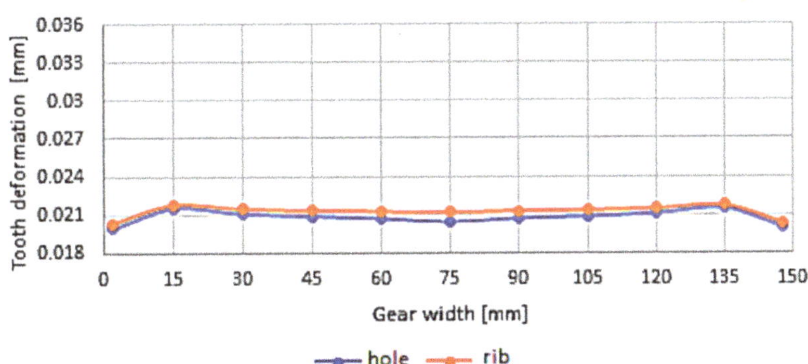

Figure 12. Deformation along the width of the tooth for model B2.

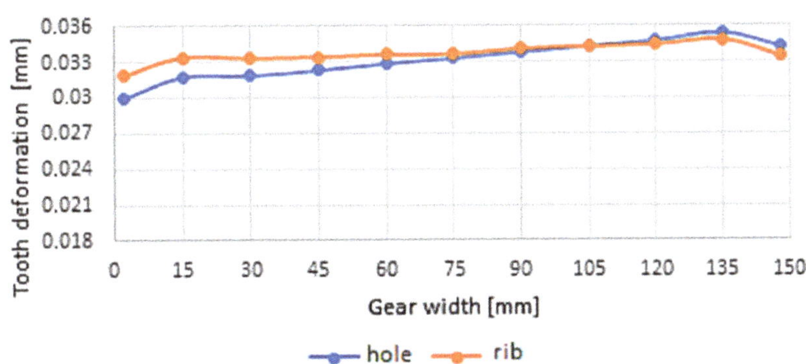

Figure 13. Deformation along the width of the tooth for model B3.

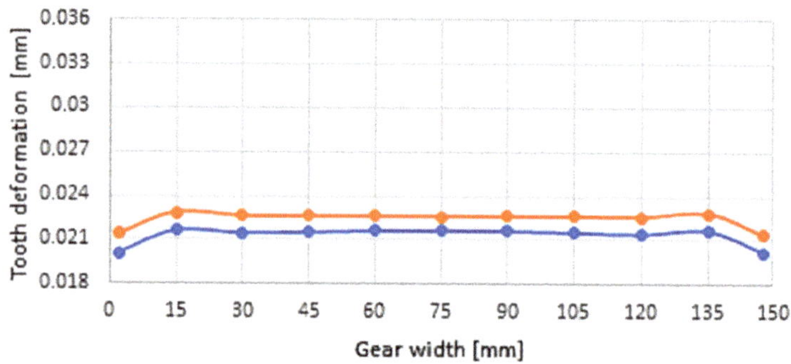

Figure 14. Deformation along the width of the tooth for model B4.

A mutual comparison of the deformation along the width of the tooth for models of cast spur gears is shown in Figure 15 if the tooth is placed above the hole, and in Figure 16 if the tooth is placed above the rib.

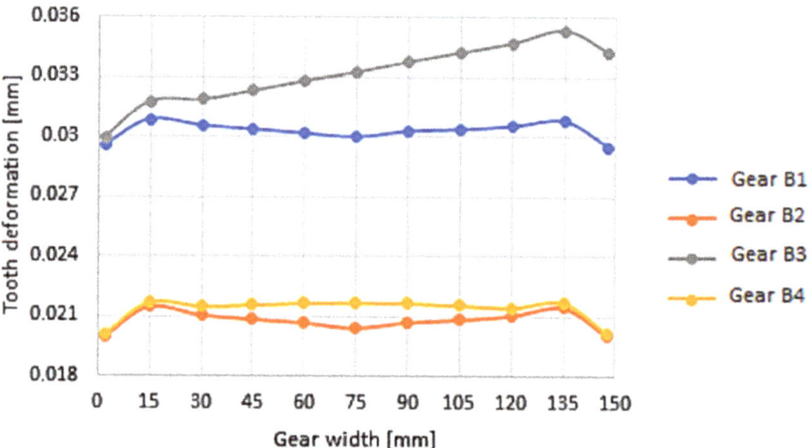

Figure 15. The deformation along the width of the tooth of the cast gear wheel located above the hole.

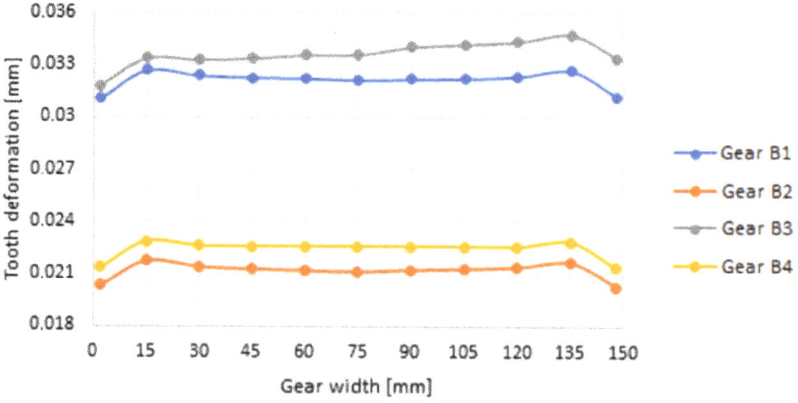

Figure 16. The deformation along the width of the tooth of the cast gear wheel located above the rib.

Figure 17 shows the stiffness of the casted gear wheel teeth which are located above the rib. The deviation of the course of deformation, as well as the stiffness of the gearing examined along the width of the gear wheel was smaller for the teeth located above the rib due to the supporting stiffness effect of the rib.

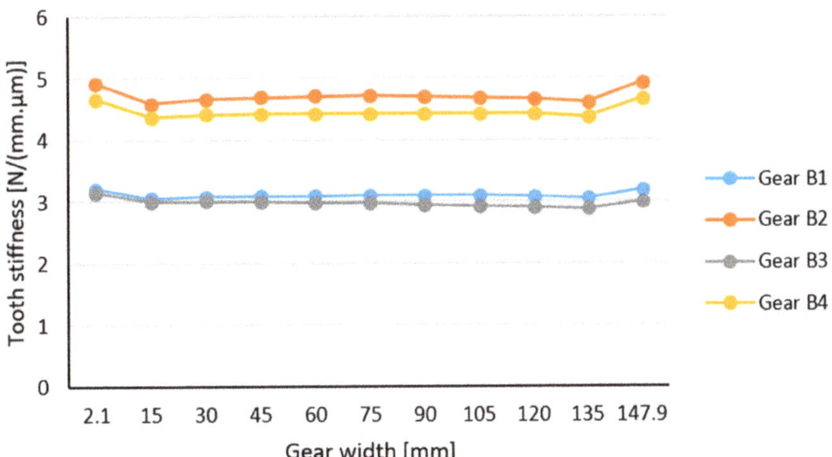

Figure 17. The stiffness along the width of the tooth of the cast gear wheel located above the hole.

Figure 18 is a comparison of the effect of weight loss on gear stiffness for cast spur gears. Wheel B4 was chosen as a wheel with zero mass loss, because its weight was the largest of the proposed models. However, the B4 wheel weighed 50.5% less than a full solid-body gear wheel. From Figure 18 it was determined that the gearing stiffness decreased with an increase in material loss. The exception was the B3 wheel due to its shape. Gear wheels B1 and B3 had the highest mass reduction, but their gear stiffnesses were smaller compared to wheels B2 and B4. The wheels B2 and B4 had almost the same amount of gear stiffness, but the wheel B2 had a greater mass reduction, which was more advantageous in terms of weight. Due to its having the highest gear stiffness and a relatively good weight, wheel B2 is the most advantageous.

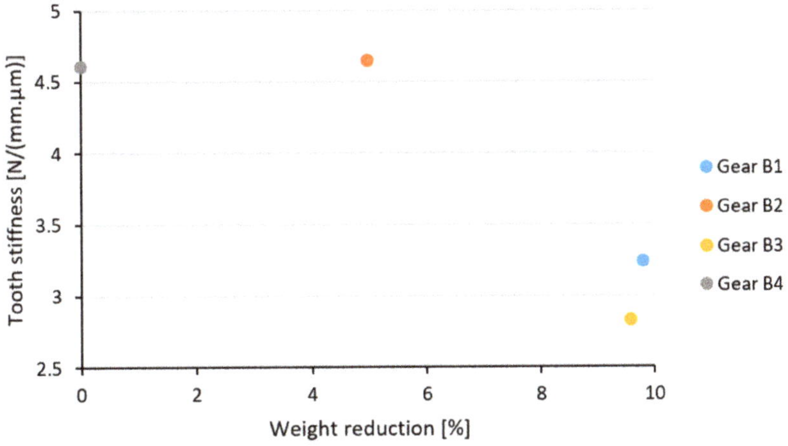

Figure 18. Influence of material loss on maximum gear stiffness for cast spur gears.

4. Conclusions

One factor in assessing the appropriate shape of a large cast spur gear wheel is to assess its suitability for deformation and the stiffness of the teeth. The steady increase in the performance factors and the decrease in the weight of the device are achieved via the development of the modern machinery and means of production. The lightening of the gear wheel body affects the deformation and stiffness of the teeth. Based on these results, it can be concluded that the fewer shape modifications made to the gear wheel body, the smaller the teeth deformation of the gear wheel.

As the thickness of the rim increases, the deformation of teeth decreases, and thus the stiffness of the teeth increases. A thickness of the rim of less than *3.5 m_n* affects the deformation of the teeth and the stiffness of the gearing more significantly. Therefore, it is advisable to choose a wreath thickness that is greater than *3.5 m_n*. The normalized module value for spur gears has the symbol m_n.

As the width of the web, which is located in the center of the width of the gear wheel, increases, the deformation of the gear teeth decreases slightly and the stiffness of the teeth increases slightly. The location of the web also affects the meshing stiffness of the gears, as do the various relief holes located in the web.

In gear design, gear noise is also a major concern. The main source of gear noise is the appearance of non-uniform rotations of the gear wheels. Periodic changes in the stiffness of the teeth during meshing in the gear are the main effects of the noise in the transfer. Reducing the dynamic loading and noise of gear systems is an important concern in gear design. When designing cast and forged gears, i.e., gears of large dimensions, it is important to assess the appropriate choice of body shape which also considers the meshing stiffness of the gearing.

Author Contributions: Conceptualization, S.M.; methodology, S.M. and M.P.; investigation, S.M.; data analysis, S.M. and S.S.; writing—original draft preparation, S.M.; writing—review and editing, M.P., S.S., P.F. and D.H. All authors have read and agreed to the published version of the manuscript.

Funding: This research is a part of these projects KEGA 029TUKE-4/2021, KEGA 027TUKE-4/2020, VEGA 1/0290/18. This work was supported by the Slovak Research and Development Agency under the Contract no. APVV-19-0328.

Institutional Review Board Statement: Not applicable.

Informed Consent Statement: Not applicable.

Data Availability Statement: The data presented in this study are part of the authors' research.

Conflicts of Interest: The authors declare no conflict of interest.

References

1. Kumar, S.M.; Govindaraj, E.; Balamurugan, D.; Daniel, F. Design analysis and fabrication of automotive transmission gearbox using hollow gears for weight reduction. *Mater. Today Proc.* **2021**, *45*, 6822–6832. [CrossRef]
2. Stroh, J.; Sediako, D. Residual Stress Characterization for Marine Gear Cases in As-Cast and T5 Heat Treated Conditions with Application of Neutron Diffraction. *Light Met.* **2019**, *2019*, 395–399.
3. Henriksson, M. Analysis of gear noise and dynamic transmission error measurements. *ASME Int. Mech. Eng. Congr. Expo.* **2004**, *47020*, 229–237.
4. Park, S.; Kim, S.; Choi, J.H. Gear fault diagnosis using transmission error and ensemble empirical mode decomposition. *Mech. Syst. Signal Process.* **2018**, *10*, 262–275. [CrossRef]
5. Dong, J.W.; Pei, W.C.; Long, H.Y.; Chu, J.; Ji, H.C. Solution of spur gear meshing stiffness and analysis of degradation characteristics. *Mechanika* **2020**, *26*, 153–160.
6. Malakova, S.; Puskar, M.; Frankovsky, P.; Sivak, S.; Palko, M.; Palko, M. Meshing Stiffness—A Parameter Affecting the Emission of Gearboxes. *Appl. Sci.* **2020**, *10*, 8678. [CrossRef]
7. Jakubovičová, L.; Ftorek, B.; Baniari, V.; Sapietová, A.; Potoček, T.; Vaško, M. Engineering Design of a Test Device. *Procedia Eng.* **2017**, *177*, 520–525. [CrossRef]

8. Wang, C. High power density design for planetary gear transmission system. *Proc. Inst. Mech. Eng. Part C J. Mech. Eng. Sci.* **2019**, *233*, 5647–5658. [CrossRef]

9. Monkova, K.; Monka, P.; Tkac, J.; Hricova, R.; Mandulak, D. Effect of the Weight reduction of a Gear Wheel on Modal Characteristics. *MATEC Web Conf.* **2019**, *299*, 1–6. [CrossRef]

10. Xu, J.; Wang, C.X. Small Module Design of Ball Mill Main Drive Gear with $\Phi 2.4 \times 10$m. *Adv. Mater. Res.* **2013**, *787*, 490–494. [CrossRef]

11. Marunić, G. Rim Stress Analysis of Thin-Rimmed Gear. *Key Eng. Mater.* **2007**, *348–349*, 141–144. [CrossRef]

12. Naveen, P.N.E.; Sujith Kumar, B.; Goriparthi, B.K.; Hota, R.S.; Chaitanya Mayee, M.; Gopala Raju, S.S.S.V. Design and analysis of thin wall gear structure with Tio2/GF reinforced Nylon66 composites. *Mater. Today Proc.* **2021**, *46*, 382–389. [CrossRef]

13. Zhao, N.; Sun, L.L.; Fu, B.B.; Li, H.F.; Wang, Q.B. Web Structural Optimization of the Big Aviation Herringbone Gear Based on APDL Language. *Appl. Mech. Mater.* **2014**, *487*, 692–698. [CrossRef]

14. Medvecká-Beňová, S. Influence of the face width and length of contact on teeth deformation and stiffness. *Sci. J. Sil. Univ. Technol. Ser. Transp.* **2016**, *91*, 99–106. [CrossRef]

15. Patel, A.; Shakya, P. Spur gear crack modelling and analysis under variable speed conditions using variational mode decomposition. *Mech. Mach. Theory* **2021**, *164*, 104357. [CrossRef]

16. Pleguezuelos, M.; Sánchez, M.B.; Pedrero, J.I. Analytical model for meshing stiffness, load sharing, and transmission error for spur gears with profile modification under non-nominal load conditions. *Appl. Math. Model.* **2021**, *97*, 344–365. [CrossRef]

17. Flek, J.; Dub, M.; Kolar, J.; Lopot, F.; Petr, K. Determination of Mesh Stiffness of Gear—Analytical Approach vs. FEM Analysis. *Appl. Sci.* **2021**, *11*, 4960. [CrossRef]

18. Xiong, Y.S.; Huang, K.; Xu, F.W.; Yi, Y.; Sang, M.; Zhai, H. Research on the Influence of Backlash on Mesh Stiffness and the Nonlinear Dynamics of Spur Gears. *Appl. Sci.* **2019**, *9*, 1029. [CrossRef]

19. Gkimisis, L.; Vasileiou, G.; Sakaridis, E.; Spitas, C.; Spitas, V. A fast, non-implicit SDOF model for spur gear dynamics. *Mech. Mach. Theory* **2021**, *160*, 104279. [CrossRef]

20. Marques, P.M.T.; Marafona, J.D.M.; Martins, R.C.; Seabra, J.H.O. A continuous analytical solution for the load sharing and friction torque of involute spur and helical gears considering a non-uniform line stiffness and line load. *Mech. Mach. Theory* **2021**, *161*, 104320. [CrossRef]

21. Shen, J.; Hu, N.Q.; Zhang, L.; Luo, P. Dynamic Analysis of Planetary Gear with Root Crack in Sun Gear Based on Improved Time-Varying Mesh Stiffness. *Appl. Sci.* **2020**, *10*, 8379. [CrossRef]

22. Císar, M.; Kuric, I.; Čuboňová, N.; Kandera, M. Design of the clamping system for the CNC machine tool. *MATEC Web Conf.* **2017**, *137*, 01003. Available online: https://www.researchgate.net/publication/321215438_Design_of_the_clamping_system_for_the_CNC_machine_tool (accessed on 20 September 2021). [CrossRef]

23. Hudak, R.; Polacek, I.; Klein, P.; Sabol, R.; Varga, R.; Zivcak, J.; Vazquez, M. Nanocrystalline Magnetic Glass-Coated Microwires Using the Effect of Superparamagnetism Are Usable as Temperature Sensors in Biomedical Applications. *IEEE Trans. Magn.* **2017**, *53*, 1–5. Available online: https://ieeexplore.ieee.org/abstract/document/7814333 (accessed on 20 September 2021). [CrossRef]

24. Jiang, H.; Sitoci-Ficici, K.; Reinshagen, C.; Molcanyi, M.; Zivcak, J.; Hudak, R.; Laube, T.; Schnabelrauch, M.; Weisser, J.; Schäfer, U.; et al. Adjustable Polyurethane Foam as Filling Material for a Novel Spondyloplasty: Biomechanics and Biocompatibility. *World Neurosurg.* **2018**, *112*, e848–e858. Available online: https://www.sciencedirect.com/science/article/abs/pii/S1878875018302171 (accessed on 20 September 2021). [CrossRef]

25. Koutecký, T.; Zikmund, T.; Glittová, D.; Paloušek, D.; Živčák, J.; Kaiser, J. X-ray micro-CT measurement of large parts at very low temperature. *Rev. Sci. Instrum.* **2017**, *88*, 033707. Available online: https://aip.scitation.org/doi/abs/10.1063/1.4979077 (accessed on 20 September 2021). [CrossRef]

26. Murčinková, Z.; Živčák, J.; Zajac, J. Experimental study of parameters influencing the damping of particulate, fibre-reinforced, hybrid, and sandwich composites. *Int. J. Mater. Res.* **2020**, *111*, 688–697. Available online: https://www.degruyter.com/document/doi/10.3139/146.111933/pdf (accessed on 20 September 2021). [CrossRef]

27. Pistek, V.; Klimes, L.; Mauder, T.; Kucera, P. Optimal design of structure in rheological models: An automotive application to dampers with high viscosity silicone fluids. *J. Vibroengineering* **2017**, *19*, 4459–4470. Available online: https://www.jvejournals.com/article/18348 (accessed on 20 September 2021). [CrossRef]

28. Puškár, M.; Jahnátek, A.; Kuric, I.; Kádárová, J.; Kopas, M.; Šoltésová, M. Complex analysis focused on influence of biodiesel and its mixture on regulated and unregulated emissions of motor vehicles with the aim to protect air quality and environment. *Air Qual. Atmos. Health* **2019**, *12*, 855–864. Available online: https://link.springer.com/article/10.1007/s11869-019-00704-w (accessed on 20 September 2021). [CrossRef]

29. Píštěk, V.; Kučera, P.; Fomin, O.; Lovska, A. Effective Mistuning Identification Method of Integrated Bladed Discs of Marine Engine Turbochargers. *J. Mar. Sci. Eng.* **2020**, *8*, 379. Available online: https://www.mdpi.com/2077-1312/8/5/379 (accessed on 20 September 2021).

30. Sabol, R.; Klein, P.; Ryba, T.; Hvizdos, L.; Varga, R.; Rovnak, M.; Sulla, I.; Mudronova, D.; Galik, J.; Polacek, I.; et al. Novel Applications of Bistable Magnetic Microwires. *Acta Phys. Pol. A* **2017**, *131*, 1150–1152. [CrossRef]

31. Sága, M.; Bulej, V.; Čuboňova, N.; Kuric, I.; Virgala, I.; Eberth, M. Case study: Performance analysis and development of robotized screwing application with integrated vision sensing system for automotive industry. *Int. J. Adv. Robot. Syst.* **2020**, *17*, 172988142092399. Available online: https://xueshu.baidu.com/usercenter/paper/show?paperid=184m0r70kg0f0a00bs7s0vq0g8225814&site=xueshu_se&hitarticle=1 (accessed on 20 September 2021). [CrossRef]
32. Toth, T.; Hudak, R.; Zivcak, J. Dimensional verification and quality control of implants produced by additive manufacturing. *Qual. Innov. Prosper.* **2015**, *19*, 9–21. Available online: https://www.qip-journal.eu/index.php/QIP/article/view/393 (accessed on 20 September 2021). [CrossRef]

Journal of
*Marine Science
and Engineering*

Analysis of Composite Scrubber with Built-In Silencer for Marine Engines

Myeong-rok Ryu and Kweonha Park *

Division of Mechanical Engineering, Korea Maritime and Ocean University, Busan 49112, Korea;
mha112@kmou.ac.kr
* Correspondence: khpark@kmou.ac.kr; Tel.: +82-51-410-4367

Abstract: The International Maritime Organization (IMO) is strengthening regulations on reducing sulfur oxide emissions, and the demand for reducing exhaust noise affecting the environment of ships is also increasing. Various technologies have been developed to satisfy these needs. In this paper, a composite scrubber for ships that can simultaneously reduce sulfur oxide and noise was proposed, and the flow characteristics and noise characteristics were analyzed. For the silencer, vane type and resonate type were applied. In the case of the vane type, the effects of the direction, size, and location of the vane were analyzed, and in the case of the resonate type, the effects of the hole location and the number of holes were analyzed. The result shows that the length increase of the vane increased the average transmission loss and had a great effect, especially in the low frequency region. The transmission loss increased when the vane was installed outside, and the noise reduction effect was excellent when the vane was in the reverse direction. In the resonate type, increasing the number of holes is advantageous for noise reduction. The condition for maximally reducing noise in the range not exceeding 840 Pa, which is 70% of the allowable back pressure, is a vane length of 225 mm in the outer vane reverse type. The pressure drop under this condition was 777 Pa, and the average transmission losses in the low frequency region and the entire frequency region were 43.5 and 54.5 dB, respectively.

Keywords: silencer; composite scrubber; transmission loss; pressure drop; noise reduction; sulfur oxides

Citation: Ryu, M.-r.; Park, K. Analysis of Composite Scrubber with Built-In Silencer for Marine Engines. *J. Mar. Sci. Eng.* **2021**, *9*, 962. https://doi.org/10.3390/jmse9090962

Academic Editor: María Isabel Lamas Galdo

Received: 17 August 2021
Accepted: 28 August 2021
Published: 3 September 2021

Publisher's Note: MDPI stays neutral with regard to jurisdictional claims in published maps and institutional affiliations.

1. Introduction

The International Maritime Organization (IMO), a specialized agency of the United Nations (UN), proposed a 0.1% sulfur content limit in fuels in the emission control areas (ECAs) from 2015, and a 0.5% sulfur content limit in areas other than ECAs from 2020 [1]. It is mandatory to use emission post-processing devices that have equal or greater effect than using low sulfur fuel [2]. Many studies have been conducted on the removal of sulfur oxides contained in exhaust gases.

Research is being conducted on wet scrubbers (applicable to existing ships). A wet scrubber, used on a ship, is a device that removes particulate matter and sulfur oxides contained in exhaust gas. It causes contact with exhaust gas through the spraying of water, removes particulate matter (capturing them in liquid droplets), and removes sulfur oxides through a chemical reaction. The effect of the shape and internal elements of the scrubber, on back pressure, particulate matter removal in exhaust gas, and chemical reaction, was studied [3–8]. Han et al. [3] compared the pressure loss and the removal efficiency of ammonia, hydrochloric acid, and hydrofluoric acid gas, according to the type of filler of the wet scrubber, and analyzed the gas removal performance by flow rate, liquid-to-liquid ratio, and pH concentration. Byeon at al. [4] designed a modified turbulent wet scrubber (MTWS) and analyzed the efficiency and pressure drop of ammonia gas and particulate removal through experiments. Lee et al. [5] analyzed the particle removal efficiency of the turbulent scrubber through experiments and studied the pressure drop to estimate the energy consumed by the scrubber. Son et al. [6] measured the smoke reduction rate before and after the scrubber by applying a wet scrubber system to a 3298cc commercial diesel engine and varying the scrubber aspect ratio, internal filling rate, washing water spray flow rate, and engine load conditions. Using this, the optimal shape and filling rates were studied. Lee et al. [7] analyzed the inlet and outlet pressure difference, internal pressure distribution, and exhaust gas streamline of the scrubber through numerical analyses to identify the problems of the existing scrubber; they designed and analyzed a new type of vortex scrubber. Hu et al. [8] studied the optimal number of blades when the dust concentration, air volume, and number of blades were different under various water intake conditions of a wet scrubber with rotating blades.

To improve the performance of the scrubber and to reduce various gases (SO_2, NO, HG^0, CO_2), various methods, such as washing water, additives, static electricity, and dust collectors were applied [9–18]. Fang et al. [9] performed basic research on the theory and technology of simultaneous SO_2 and NO treatments by spraying urea water. Fang et al. [10] conducted a study on the simultaneous treatment of SO_2, NO, and HG^0 by spraying urea water and $KMnO_4$. Raghunath and Mondal [11] conducted a study on the simultaneous reduction of SO_2 and NO by mixing and spraying $NH_3/NaClO$ with water. Yang et al. [12] studied the effect of electrolyzed seawater injection on NO_x and SO_2 reduction. Meikpa et al. [13] studied the SO_2 removal efficiency of a multi-stage bubble column scrubber using water. D'Addio et al. [14] conducted a study on the removal of submicron particles using wet electrostatic scrubbing. Park et al. [15] studied the simultaneous reduction of NO and SO_2 using a wet scrubber combined with a plasma electrostatic precipitator. Lim et al. [16] investigated the contamination characteristics of the washing water at the scrubber and studied the removal characteristics of sulfur compounds when microbubbles were supplied to the washing water. Sun et al. [17] conducted a study to simultaneously reduce CO_2 and PM in industrial flue gas by connecting an ammonia scrubber (AS) and a granular bed filter (GBF) in series. Xie et al. [18] conducted a study on VOCs (Volatile Organic Compounds) removal of a wet scrubber combined with UV/PMS (peroxymonosulfate).

There are no regulations on the exhaust noise from ships, but noise regulation is being considered, regarding the impact on the health of the crew and the surrounding residential areas when anchored at the port. Diesel engines are the main sources of vibration noise for ships because they generate vibration noise through combustion, piston up-and-down motions, and a crankshaft rotational motion in the process of producing mechanical energy [19]. Since the engine is inside the ship, the engine noise spreading to the inside of the ship is greatly reduced by the partition walls and structures. However, since engine noise transmitted through the exhaust gas is hardly reduced, the noise of the ship can be greatly reduced by installing a silencer at the exhaust gas outlet.

Noise reduction is caused by the synthesis, dissipation, and conversion of sound waves into thermal energy. The silencer is divided into a resonance type that focuses on sound wave synthesis, a vane type that induces dissipation and synthesis at the same time, and an expansion type that focuses on dissipation. Various studies are being conducted on the flow characteristics and acoustic characteristics according to the shape of the silencer for noise reduction [20–27]. Hwang et al. [20] investigated the internal structure of the silencer, classified its components, and studied the noise and back pressure characteristics of the silencer. Kwon et al. [21] assumed that the fluid flow in the simple expansion chamber was divided into potential flow and turbulent flow, and then studied the acoustic transmission loss of the muffler according to the flow velocity of the fluid through computational analysis. Kim and Jeong [22] modeled a circular simple expansion pipe divided by a bulkhead with holes, and studied acoustic and flow characteristics through computational analysis. Yi et al. [23] studied the pressure loss through the internal flow analysis of a silencer for an 8 kW diesel generator. Secgin et al. [24] studied the acoustic characteristics through computational analysis by changing the baffle position, number, and extension shape of the baffle in the expansion tube. Nursal et al. [25] studied the flow characteristics according to the pipe shape and the position of the silencer of a 4-stroke marine diesel generator through computational analysis. Kakade and Sayyad [26] studied the transmission loss and back pressure according to the shape of the silencer for automobiles through computational analysis. Kim [27] studied the structure of an effective silencer for high-frequency attenuation while maintaining the performance of the existing silencer for medium and low frequency attenuation in exhaust noise generated from medium-sized diesel engines.

In this paper, a composite scrubber that can simultaneously reduce exhaust gas and engine noise was proposed, and the factors affecting noise reduction and the effect on engine performance were analyzed. Among the types of silencers currently used in various fields, vane-type and resonance-type silencers are combined with the scrubber. The effect of the length change, the installation position and the inclined direction of the vane on the flow characteristics and acoustic characteristics was studied in the vane type silencer, and the effect of the number and installation position of the holes was analyzed in the resonator type silencer. Through this analysis, a design method for maximally reducing noise in the range of 840 Pa or less, which is 70% of the allowable back pressure of the target engine, is suggested.

2. Mathematical Model and Calculation Conditions

2.1. Mathematical Model

2.1.1. CFD Mathematical Models

The computational analysis program used to analyze the flow characteristics is ANSYS CFX V14.0 (Ansys Inc.: Canonsburg, PA, USA) and the formulas refer to the ANSYS CFX-Theory Guide [28].

The continuity equation is given in Equation (1).

$$\frac{\partial \rho}{\partial t} + \nabla \cdot (\rho U) = 0 \tag{1}$$

The momentum is given in Equation (2).

$$\frac{\partial (\rho U)}{\partial t} + \nabla \cdot \left(\rho U \otimes U \right) = -\nabla p + \nabla \cdot \tau + S_M \tag{2}$$

where, stress tensor τ is associated with strain rate, as shown in Equation (3).

$$\tau = \mu \left[\nabla U + (\nabla U)^T - \frac{2}{3} \delta \nabla \cdot U \right] \tag{3}$$

The total Energy Equations is given in Equation (4).

$$\frac{\partial (\rho h_{tot})}{\partial t} - \frac{\partial p}{\partial t} + \nabla \cdot (\rho U h_{tot}) = \nabla \cdot (\lambda \nabla T) + \nabla \cdot (U \cdot \tau) + U \cdot S_M + S_E \tag{4}$$

where total enthalpy h_{tot} is associated with static enthalpy $h(T, P)$ as shown in Equation (5).

$$h_{tot} = h + \frac{1}{2} U^2 \tag{5}$$

$\nabla \cdot (U \cdot \tau)$ is the term for work by viscous stress and indicates internal heating by viscosity of fluid.

2.1.2. Acoustic Mathematical Model

The computational analysis program used to analyze the acoustic characteristics is ANSYS Acoustics ACT (Application Customization ToolKit) Extension (Ansys Inc.: Canonsburg, PA, USA), and it was installed and used in the Harmonic response program of ANSYS Workbench V17.0 (Ansys Inc.: Canonsburg, PA, USA). For the formula, refer to Lectures of the Acoustics ACT Extension [29].

In acoustic fluid-structural interaction problems, the structural dynamics equation must be considered along with the Navier-Stokes equations of fluid momentum and the flow continuity equation. The fluid momentum equation and continuity equation are simplified to obtain the acoustic wave equation using the assumption that the fluid is compressible and has no mean flow. That equation is given in Equation (6).

$$\nabla \cdot \left(\frac{1}{\rho_0}\nabla p\right) - \frac{1}{\rho_0 c^2}\frac{\partial^2 p}{\partial t^2} + \nabla \cdot \left[\frac{4\mu}{3\rho_0}\nabla\left(\frac{1}{\rho_0 c^2}\frac{\partial p}{\partial t}\right)\right] = -\frac{\partial}{\partial t}\left(\frac{Q}{\rho_0}\right) + \nabla \cdot [\frac{4\mu}{3\rho_0}\nabla\left(\frac{Q}{\rho_0}\right) \tag{6}$$

where, c is speed of sound ($\sqrt{\frac{K}{\rho_0}}$) in fluid medium, ρ_0 is mean fluid density, μ is dynamic viscosity, K is bulk modulus of fluid, p is acoustic pressure, Q is mass source in the continuity equation and t is time.

The acoustic pressure exerted on the structure at the FSI (Fluid Structure Interaction) interface will be considered in the Derivation of Acoustics matrices to form the coupling stiffness matrix. Harmonically varying pressure is given in Equation (7)

$$p(r,t) = Re\left[p(r)e^{jwt}\right] \tag{7}$$

where, p is amplitude of the pressure, j is $\sqrt{-1}$, w is 2nf and f is frequency of oscillations of the pressure.

The transmission losses is given in Equation (8).

$$TL = 10\log\left(\frac{W_{in}}{W_{transmitted}}\right)(\text{dB}) \tag{8}$$

where, W_{in} represents the radiated sound pressure, $W_{transmitted}$ represents the transmitted sound pressure.

The acoustic sound pressure level is calculated as Equation (9).

$$L_{SPL} = 10\log\left(\frac{P_s^2}{P_{ref}^2}\right)(\text{dB}) \tag{9}$$

P_s is the root mean square of sound pressure, and P_{ref} is the reference sound pressure that is $2 \times 10^{-5}(\text{Pa})$.

2.2. Calculation Grids

2.2.1. Flow Calculation

Figure 1 is a mesh for flow analysis created using ICEM CFD (Ansys Inc.: Canonsburg, PA, USA). The complex shape was composed of a Tetra mesh, and the number of meshes was set to 1.2 million, 1.7 million, 2.3 million, and 3.3 million, and the effect of the number of meshes on the analysis was confirmed. Figure 2 shows the pressure drop according to the number of meshes. The effect of the number of meshes on the pressure drop was small and was stabilized from 3 million pieces. In this study, flow analysis was performed by selecting the number of meshes of 3.3 million in consideration of the mesh quality of complex shapes.

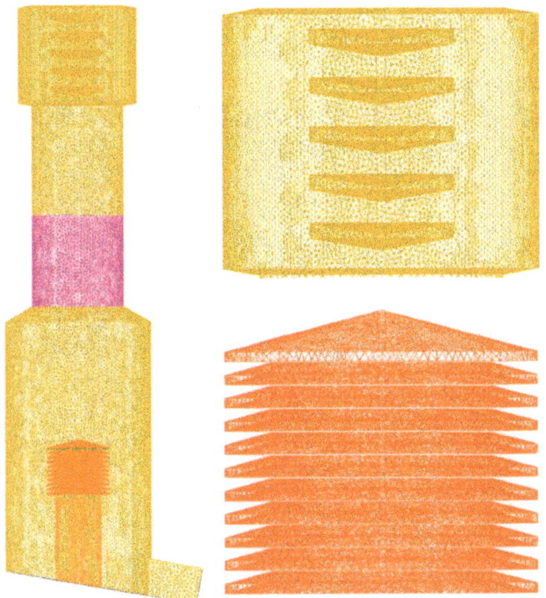

Figure 1. Mesh shape for flow calculation.

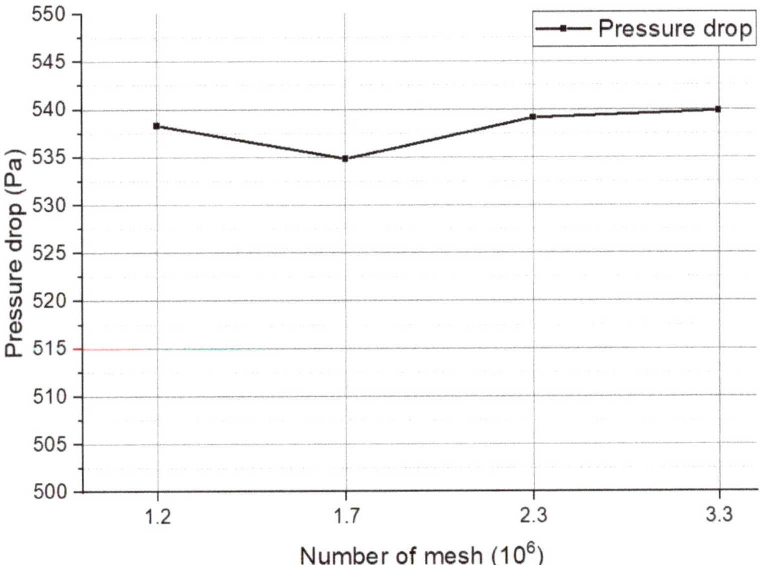

Figure 2. Pressure drop with the number of grids.

2.2.2. Acoustics Calculation

Figure 3 is a mesh for acoustic analysis created using ANSYS Mechanical mesh; 1/2 symmetry condition was used and it was created as a hexa mesh. The number of meshes was increased by 50,000 from 50,000 to 500,000 and the effect on the analysis results was evaluated. Figure 4 shows the average transmission loss from 100 to 2000 Hz, according to the number of meshes. As a result of analyzing the effect of the mesh, it was confirmed that the average transmission loss converges from 350,000 pieces, and this mesh was applied to the acoustics calculation.

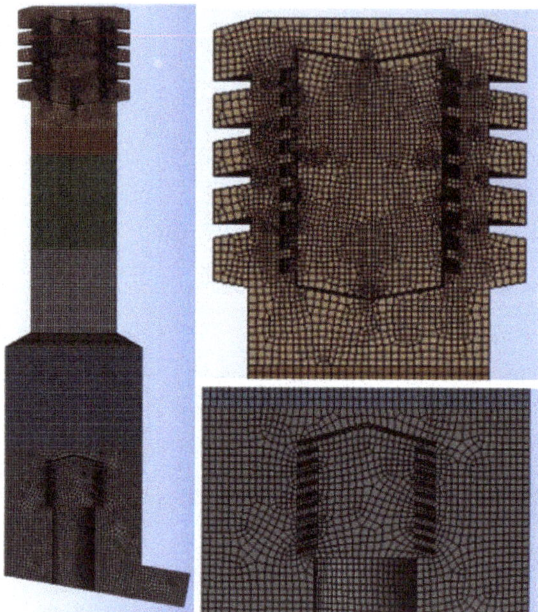

Figure 3. Mesh for acoustics calculation.

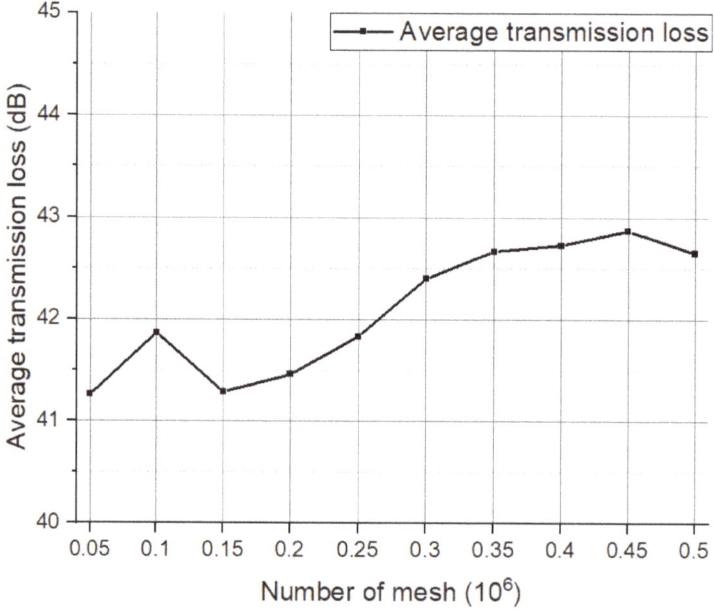

Figure 4. Transmission loss with the number of grids.

2.3. Calculation Reliability

2.3.1. Flow Calculation

The ANSYS CFX program used to analyze the flow was used for various flow analysis as a commercial code and had sufficient reliability. However, for the porous condition that simulates the filling layer, the conditions must be set according to the filler used in this study, and it is necessary to compare it with the performance of the actual product. The filler used in this study is a plastic pall ring. Porous conditions were set using the pressure drop values of one-inch pall rings given in [30] (p. 12), and the pressure drop values were compared under the same conditions. Figure 5 compares the simulated value and the experimental value of the pressure drop. The comparison shows that the results are similar values in the overall area.

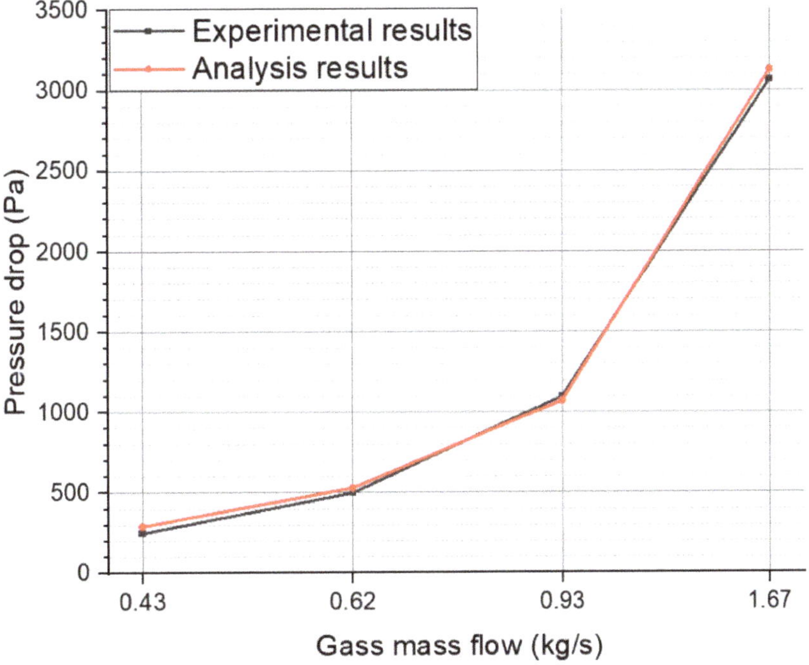

Figure 5. Comparison of experimental and analysis pressure drop results.

2.3.2. Acoustics Calculation

Acoustic behavior was analyzed using Acoustics ACT provided by ANSYS. In order to secure the reliability of the acoustic analysis, test report through experiments were requested. Moreover, the experimental and analysis values were compared. The sound loss was measured by generating a sound source inside the reverberation room and calculating the difference between the sound pressure level L1 (dB) measured 1 m in front of the speaker and the sound pressure level L2 (dB) measured after passing through the scrubber. Figure 6 is a graph comparing the experimental and analysis values. As a result of comparing the experimental and analysis values, the error was less than 2 dB.

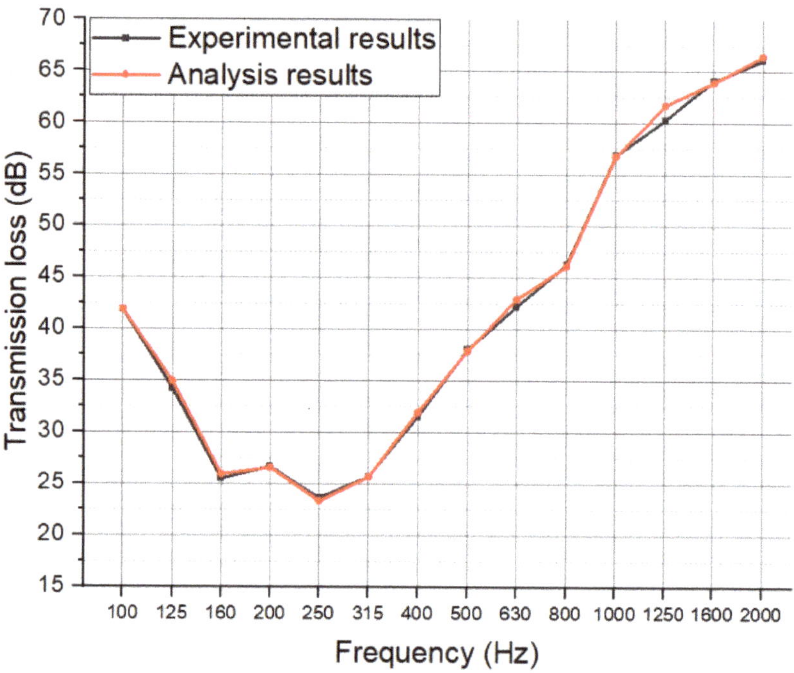

Figure 6. Comparison of experimental results and simulation results.

2.4. Calculation Condition

Figure 7 shows the structure and cross-sectional view of the composite scrubber. The structure of the composite scrubber is divided into four parts: inlet, spray, porous, and silencer. The diameter of the inlet area is 800 mm, the diameter of the porous area and the spray area is 500 mm, and the diameter of the silencer area is 650 mm. The heights of the inlet, porous, spray, and silencer area are 1500, 500, 600, and 600 mm, respectively. Five sprays are located 400 mm above the bottom of the spray domain and are 150 mm apart from the center in four directions: top, bottom, and both sides. The bottom of the inlet area was inclined by 10° to facilitate the discharge of washing water. A guide vane was installed above the inlet to prevent the sprayed water from flowing back into the inlet. The inclination of the vanes was 10°, the number of vanes was 11, and the distance between the vanes was 10 mm.

Table 1 shows the test cases. Vane type and resonate type were applied to the scrubber. Vane surface was covered with a sound-absorbing material of polyether 24. The length and inclination direction of the vane had a great effect on the sound waves due to the phase change of the sound waves, and had a great influence on the sound absorption performance, according to the area variation of the sound absorption material. Therefore, the effect was analyzed by increasing the length of the vane by 25 mm, from 150 mm to 250 mm, and the effect was analyzed by changing the inclination direction angle of the vane by 10° in the forward and reverse directions to the flow direction of the fluid. In the Resonate type, the same sound absorbing material was attached to the inner wall of the resonate tube, and the hole location and number of holes were expected to affect the sound wave synthesis. The effect was analyzed by increasing the number of holes by 30 from 60 to 150.

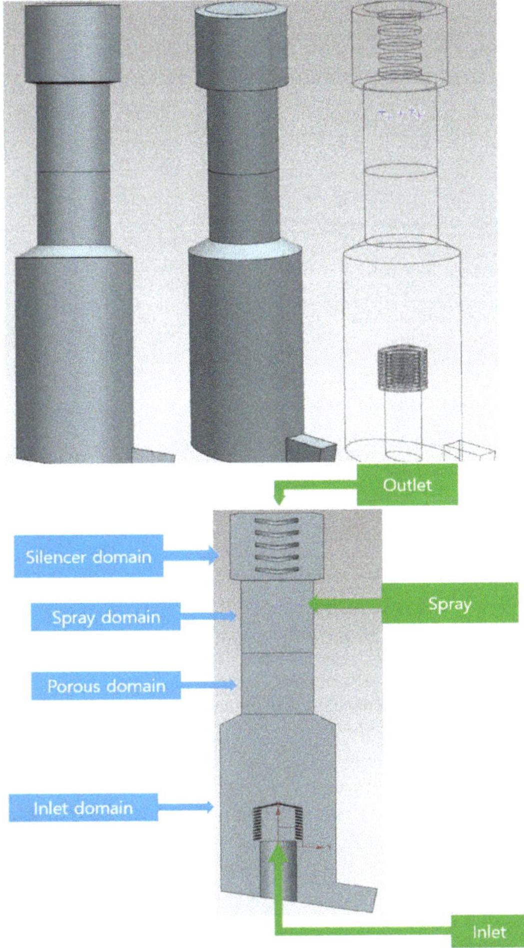

Figure 7. Overall shape and sectional view of composite scrubber.

Table 1. Calculation cases.

Cases	Type	Variable		Shape
Case 1			150	
Case 2			175	
Case 3	Inner vane type	Vane length (mm)	200	
Case 4			225	
Case 5			250	
Case 6			150	
Case 7			175	
Case 8	Inner vane reverse type	Vane length (mm)	200	
Case 9			225	
Case 10			250	
Case 11			150	
Case 12			175	
Case 13	Outer vane type	Vane length (mm)	200	
Case 14			225	
Case 15			250	

Table 1. *Cont.*

Cases	Type	Variable		Shape
Case 16			150	
Case 17			175	
Case 18	Outer vane reverse type	Vane length (mm)	200	
Case 19			225	
Case 20			250	
Case 21			60	
Case 22	Inner resonate type	Number of holes	90	
Case 23			120	
Case 24			150	
Case 25			60	
Case 26	Outer resonate type	Number of holes	90	
Case 27			120	
Case 28			150	

Table 2 shows the specifications of the target engine used in the study. Allowable pressure means the limit of engine back pressure (it was 1200 Pa).

Table 2. Target engine specification.

Engine Model	YANMAR 6-MAL
Number of cylinder	4
Bore × stroke (mm)	200 × 240
Combustion chamber type	Direct injection
Rated output (hp)	640
Rated speed (rpm)	900
Allowable pressure (Pa)	1200

Table 3 shows the calculation conditions used for flow analysis. Applied materials are ideal air, water liquid, and water vapor, and the phase change between water liquid and vapor was considered. The number of sprays was five, at a flow rate of 1.1 kg/s. The gas inlet flow rate was 0.55 kg/s and the temperature was 230 °C. The outlet was set equal to atmospheric pressure, and the porosity of the porous domain was set to 0.902. The server's CPU (Central Processing Unit) used for analysis is Intel® Xeon® 2nd generation (Intel corporation: Santa Clara, CA, USA), and the analysis time per each case was 10 to 12 h.

The pressure drop represents the difference in pressure between the inlet and outlet. Since the pressure is fixed at the atmospheric pressure at the outlet, the pressure generated by the inlet minus the atmospheric pressure becomes the pressure drop, which is the same as the engine back pressure. Considering various devices, such as the turbocharger and economizer mounted after the engine exhaust, the back pressure generated by the composite scrubber should not exceed 840 Pa, which is 70% of the allowable engine pressure. If it exceeds this value, the case is excluded from the optimal shape.

In the case of acoustic analysis, the inlet frequency condition is given as 100, 125, 160, 200, 250, 315, 400, 500, 630, 800, 1000, 1250, 1600, 2000 Hz. The sound absorption coefficient of polyester (24 kg/m^3 25 mm) used for the inner wall of vane and resonance tube was given by [31] (p. 104).

Table 3. Calculation conditions.

Analysis Type		Steady-State
Turbulence model		SST (Shear Stress Transport)
Constituent material		Air, water vapor, water liquid
Wall condensation model		Concentration boundary layer model
H_2O/H_2O_l component pair details		Liquid evaporation model
Particle injection	Cone angle (°)	30
	Particle diameter (mm)	2
	Particle mass flow rate (kg/s)	1.1
	The number of injection Regions	5
Initial conditions	Static pressure (atm)	1
	Temperature (°C)	30
Air inlet	Mass flow rate (kg/s)	0.55
	Static temperature (°C)	230
Air outlet	Opening pressure (atm)	1
Porous domain	Volume porosity	0.902

3. Result and Discussion

3.1. Inner Vane Type

Figure 8 shows the transmission loss in the entire frequency domain, according to the increase in the length of the inner vane type. At frequencies below 160 Hz, the effect of the vane length on transmission loss was small, and at frequencies above 160 Hz, transmission loss increased as the length increased. In the case of the inner vane type, the transmission loss decreased at 100–160 Hz, 500–750 Hz and 1250–1600 Hz, and the transmission loss increased at 315–500 Hz, 800–1250 Hz and 1600–2000 Hz.

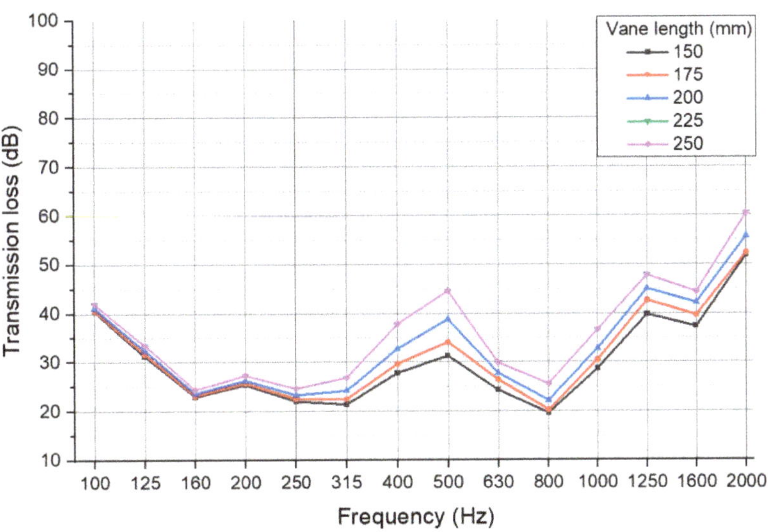

Figure 8. Transmission loss according to the vane length in the inner vane type.

Figure 9 shows the average transmission loss in the low frequency range of 100–200 Hz and the entire frequency range of 100–2000 Hz, according to the change in the length of the vane. The average transmission loss in the low frequency region increased a little, but the transmission loss increased as the length increased. However, there was almost no difference in transmission loss over 225 mm. The average transmission loss at all the frequencies also showed the same trend and the increase was larger. The average transmission loss in the entire frequency region showed maximum values of 36.1 dB.

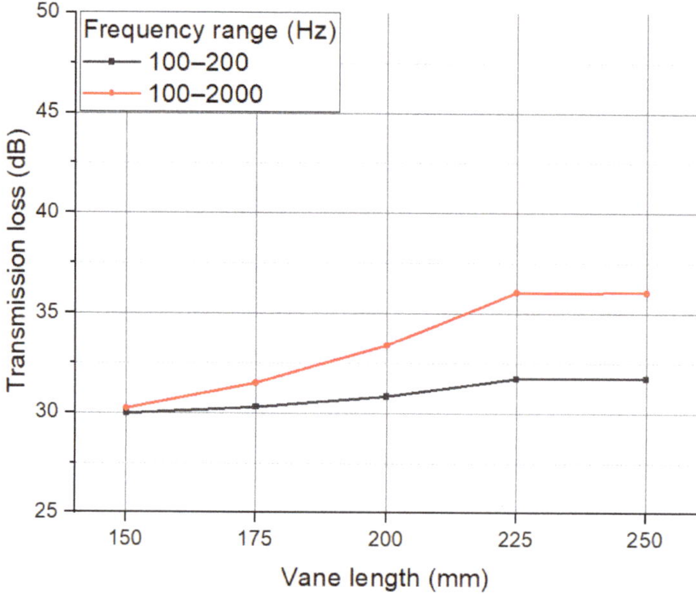

Figure 9. Average transmission loss according to the vane length in the inner vane type.

Figures 10 and 11 show the pressure distribution and pressure drop of the scrubber according to the vane length in the inner vane type. As the length of the vane increases, the pressure inside the scrubber increases and the pressure drop increases. As the length increases, the pressure drop increases rapidly. The pressure drop increases from 537 to 648 Pa.

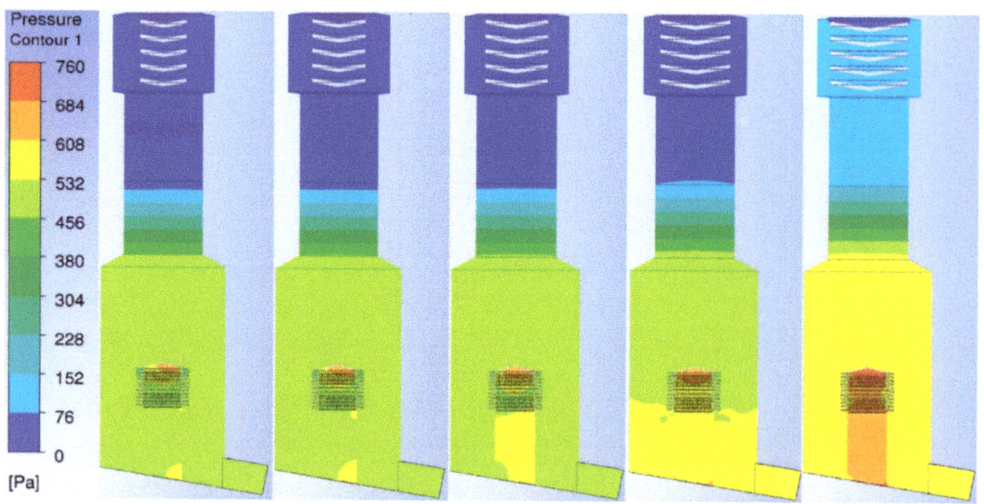

Figure 10. Pressure contour according to the vane length in the inner vane type.

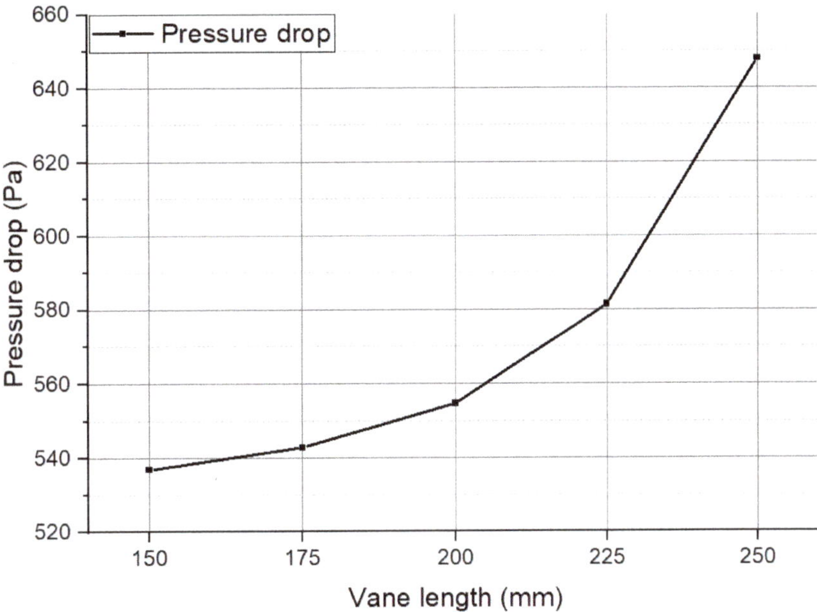

Figure 11. Pressure drop according to the vane length in the inner vane type.

In the inner vane type, increasing the vane length increases the transmission loss and pressure drop. However, after 225 mm, there is no effect on the transmission loss and only the pressure drop value increases significantly.

3.2. Inner Vane Reverse Type

Figure 12 shows the transmission loss according to the increase in the length of the vane in the inner vane reverse type. It has similar characteristics and trends to the inner vane type.

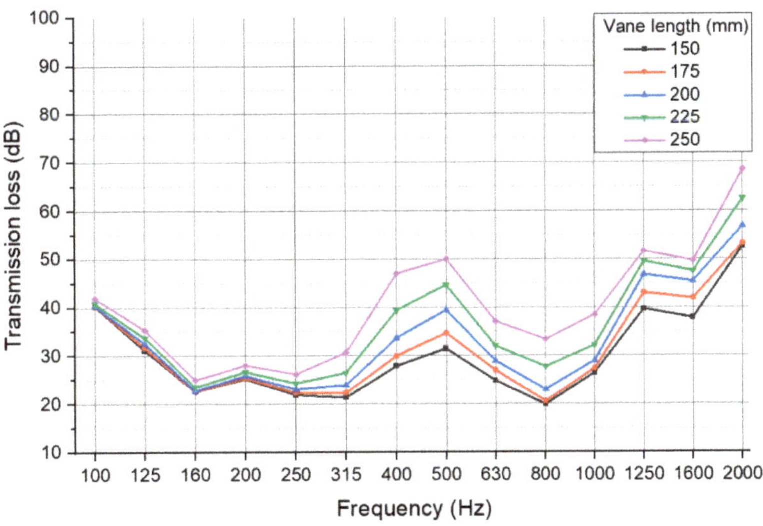

Figure 12. Transmission loss according to the vane length in the inner vane reverse type.

Figure 13 shows the average transmission loss in the low frequency and the entire frequency range according to the vane length. The transmission loss in the low frequency region and entire frequency region showed a similar shape to that of the inner vane type, but the difference was further increased at the length of 250 mm. Transmission losses in the low frequency and whole frequency domains were 32.6 and 40.2 dB at 250 mm, respectively.

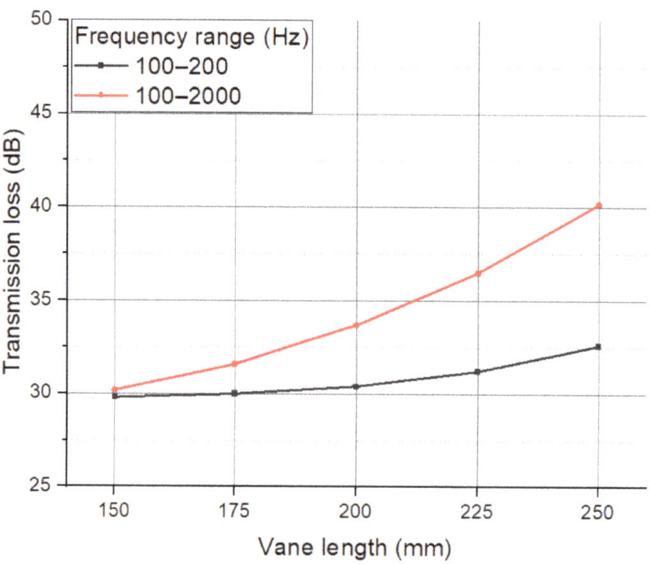

Figure 13. Average transmission loss according to the vane length in the inner vane reverse type.

Figures 14 and 15 show the pressure contour and pressure drop of the scrubber according to the length of the vane in the inner vane reverse type. The pressure change inside the scrubber is similar to the inner vane type. The pressure drop increases from 535 to 650 Pa with increasing length. In particular, it increases rapidly when it exceeds 225 mm.

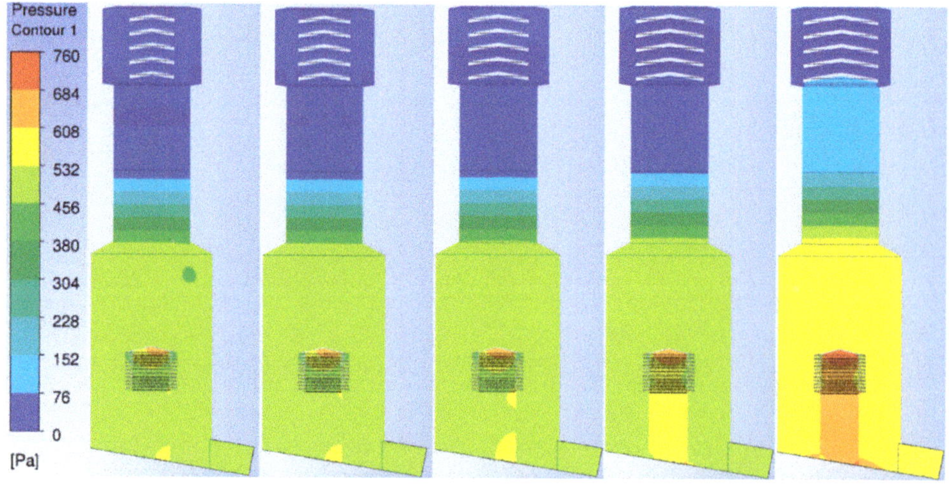

Figure 14. Pressure contour according to the vane length in the inner vane reverse type.

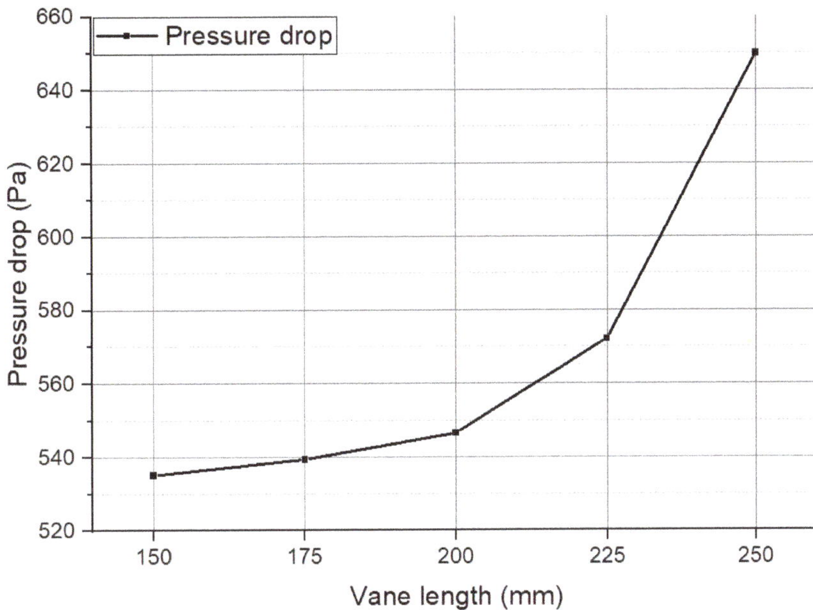

Figure 15. Pressure drop according to the vane length in the inner vane reverse type.

3.3. Outer Vane Type

Figure 16 shows the transmission loss in the entire frequency range according to the increase of the vane length in the outer vane type. Unlike the inner vane type, the transmission loss increases as the length of the vane increases at all frequencies. At 100–160 Hz and 500–630 Hz, it shows a clear decrease regardless of the length, and it shows an increase in the 200–250 Hz region. The transmission loss increases as the length increase, and the maximum loss appears in the region of 250–400 Hz. The increase and decrease are repeated in the region after 750 Hz, but the overall transmission loss tends to increase.

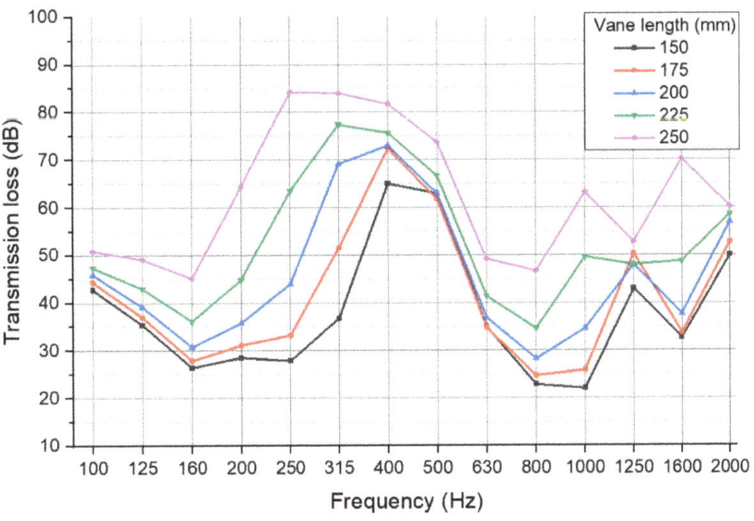

Figure 16. Transmission loss according to the vane length in the outer vane type.

Figure 17 shows the transmission loss in the low and entire frequency regions, according to the change in the length of the vane. In the low frequency and entire frequency regions, the transmission loss increases as the length increases. The maximum transmission losses in the low frequency region and in the entire frequency region are 52.3 and 62.5 dB at 250 mm, respectively.

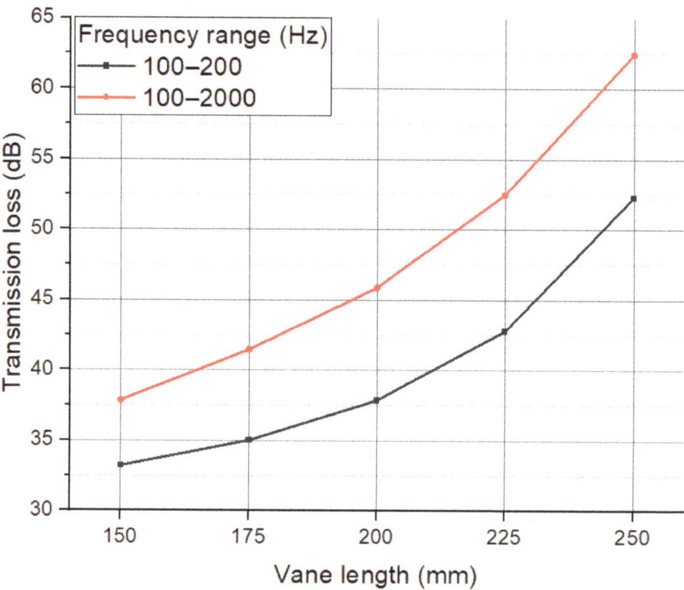

Figure 17. Average transmission loss according to the vane length in the outer vane type.

Figures 18 and 19 show the pressure contour and pressure drop of the scrubber according to the length of the vane in the outer vane type. The red line in Figure 19 represents 840 Pa, which is 70% of the allowable engine pressure. As the length of the vane increases, the pressure inside the scrubber increases and the pressure drop increases. As the length increases, the pressure drop increases and it increases significantly after 225 mm. When increasing from 150 to 250 mm, the pressure drop increases from 551 to 1186 Pa.

Figure 18. Pressure contour according to the vane length in the outer vane type.

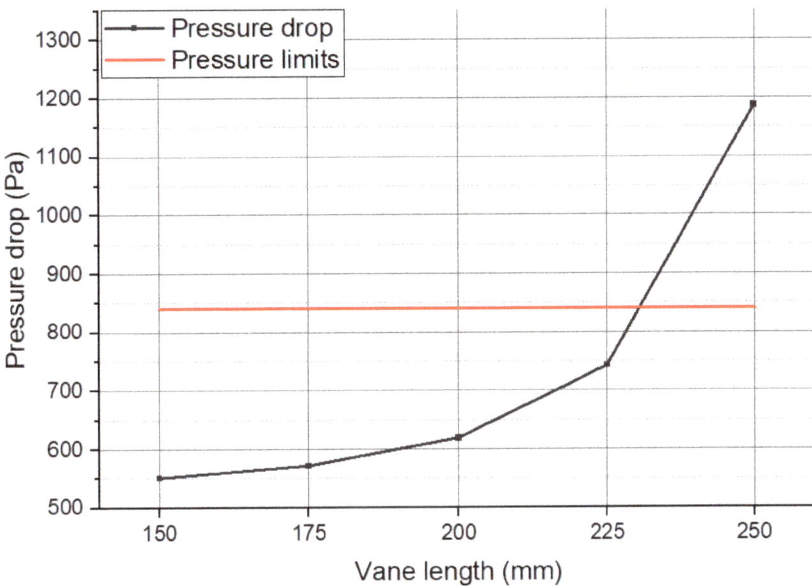

Figure 19. Pressure drop according to the vane length in the outer vane type.

In the outer vane type, increasing the length of the vane greatly increases the transmission loss and pressure drop, and the pressure drop at 250 mm greatly exceeds 840 Pa, which may have a negative effect on the engine.

3.4. Outer Vane Reverse Type

Figure 20 shows the transmission loss in the entire frequency range as the length of the vane increases in the outer vane reverse type. It has similar characteristics and tendencies to the outer vane type, but overall transmission loss is high.

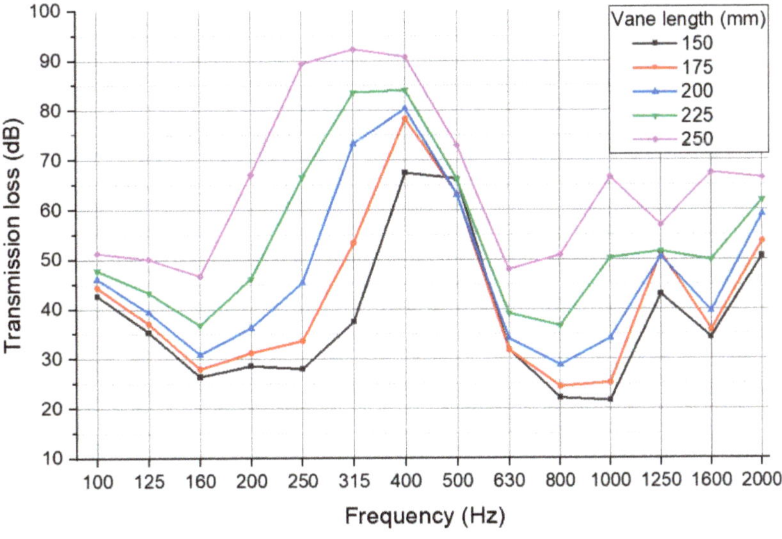

Figure 20. Transmission loss according to the vane length in the outer vane reverse type.

Figure 21 shows the transmission loss in the low frequency and the entire frequency regions according to the change in the length of the vane. It has a similar tendency to the outer vane type, and the increase in transmission loss is higher. The maximum transmission loss is 65.4 dB at 250 mm.

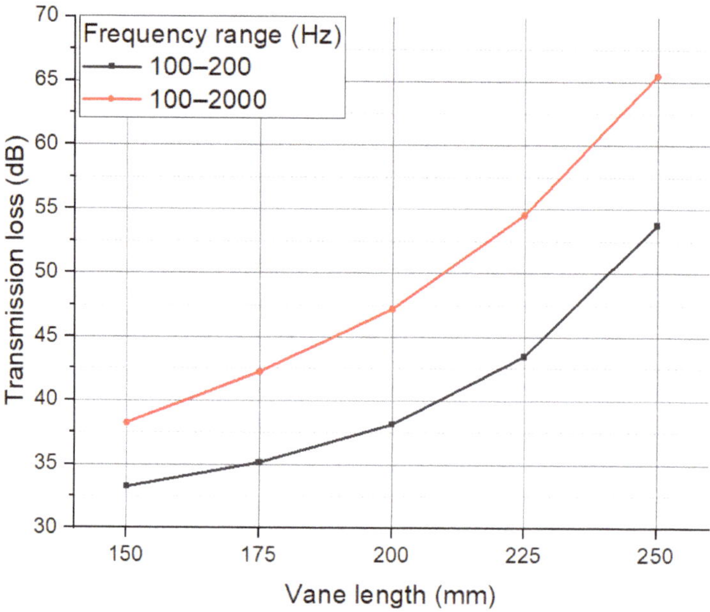

Figure 21. Average transmission loss according to the vane length in the outer vane reverse type.

Figures 22 and 23 shows the pressure contour and pressure drop of the scrubber according to the length of the vane in the outer vane reverse type. It shows a similar tendency and characteristics to the outer vane type, but the increase in pressure drop is higher. The pressure drop increases from 555 to 1286 Pa.

Figure 22. Pressure contour according to the vane length in the outer vane reverse type.

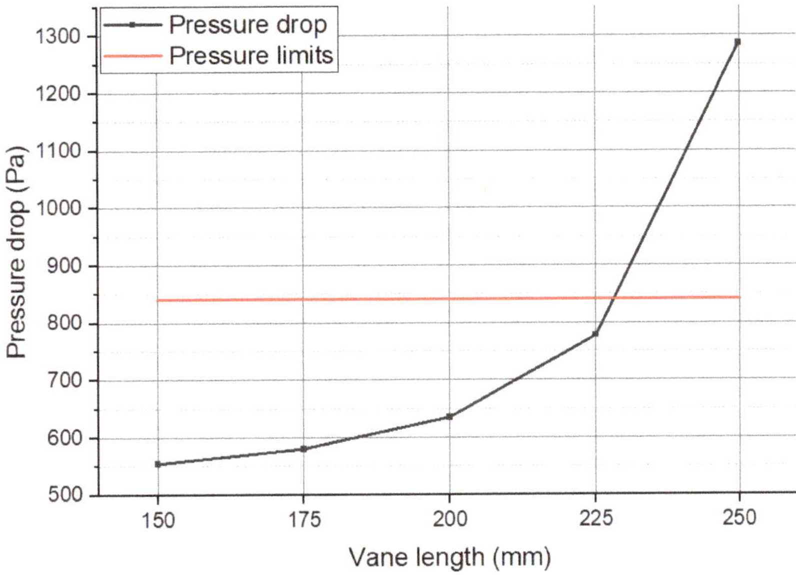

Figure 23. Pressure drop graph according to the vane length in the outer vane reverse type.

In the outer vane reverse type, increasing the length of the vane greatly increases both the transmission loss and pressure drop. At 250 mm, the pressure drop value greatly exceeds the allowable engine value of 840 Pa, which can negatively affect the engine.

3.5. Inner Resonate Type

Figure 24 shows the transmission loss in the entire frequency domain according to the number of holes in the inner resonate type. Except for the 200–500 Hz and 2000 Hz regions, the effect of the number of holes on the transmission loss was not much. Transmission loss decreased in the range of 100–160 Hz, 400–800 Hz, and increased in other regions.

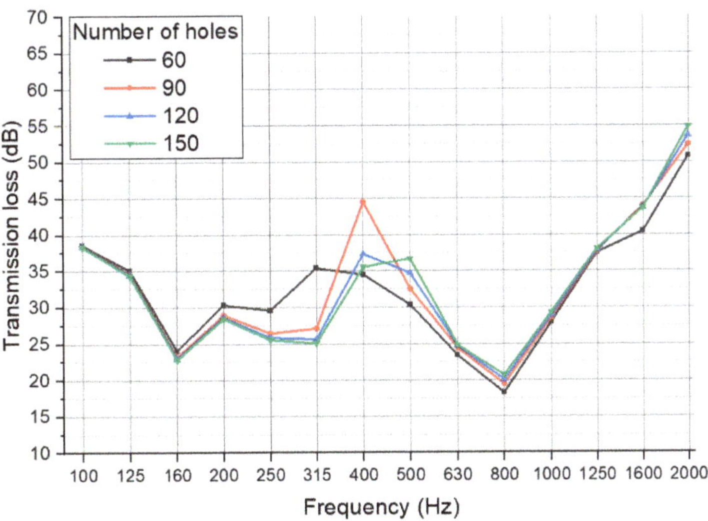

Figure 24. Transmission loss according to the number of holes in the inner resonate type.

Figure 25 shows the average transmission loss in the low frequency and the entire frequency regions according to the change in the number of holes. The average transmission loss in the low frequency region tends to decrease as the number of holes increases, but the average transmission loss in the entire frequency region repeats the increase and decrease and is not significantly affected by the number of holes. The maximum transmission loss was 32 dB at 60 holes in the low frequency region, and 33 dB at 90 holes in the entire frequency region.

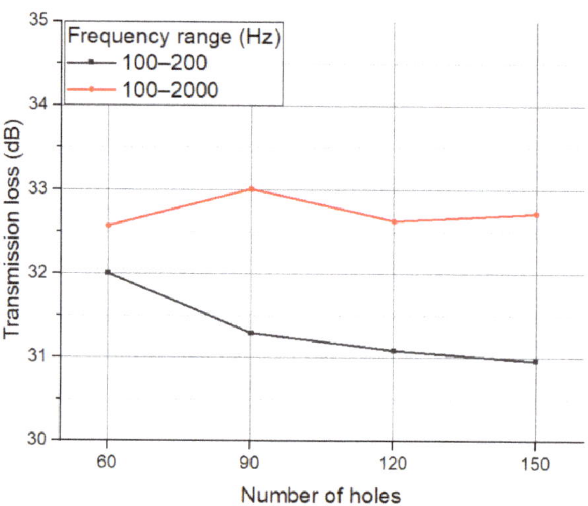

Figure 25. Average transmission loss according to the number of holes in the inner resonate type.

Figures 26 and 27 show the pressure contour and pressure drop of the scrubber according to the change in the number of holes in the inner resonate type. The number of holes does not have much of an effect on the pressure inside the scrubber, and the pressure drop does not show a big difference. When increasing from 60 to 150 holes, the pressure drop increases from 601.6 to 602.8 Pa.

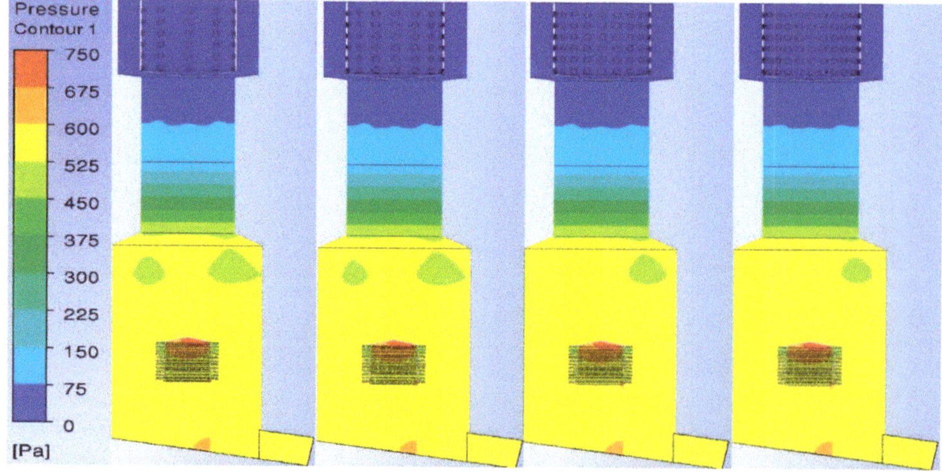

Figure 26. Pressure contour according to the number of holes in the inner resonate type.

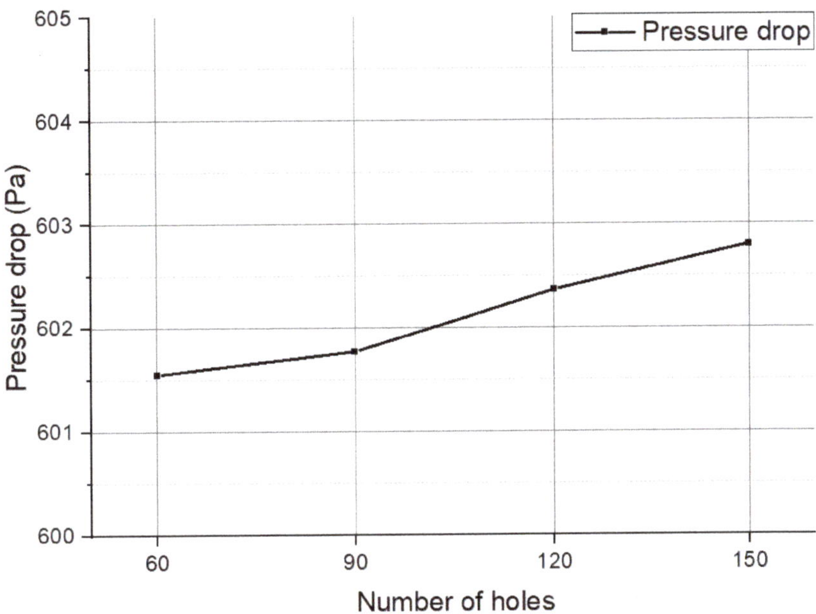

Figure 27. Pressure drop graph according to the number of holes in the inner resonate type.

In the inner resonate type, as the number of holes increases, the average transmission loss in the entire frequency region is not that different, and the pressure drop increases slightly from 601.6 to 602.8 Pa.

3.6. Outer Resonate Type

Figure 28 shows the transmission loss in the entire frequency domain according to the increase in the number of holes in the outer resonate type. Except for the 200–800 Hz region, the effect of the number of holes on the transmission loss was not much. The transmission loss shows a decrease in the range of 100–160 Hz, 400–800 Hz, and tends to rise in the other ranges.

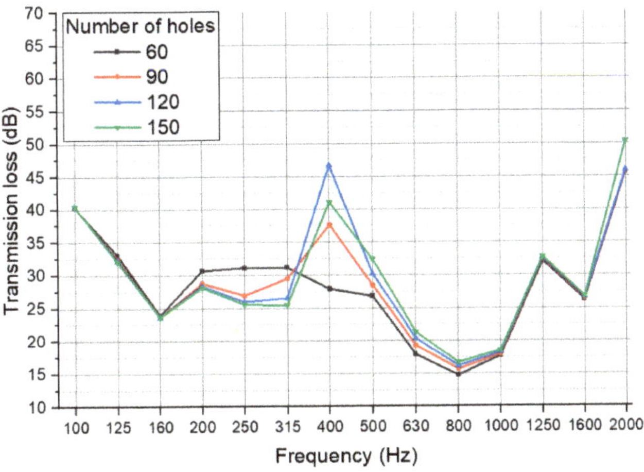

Figure 28. Transmission loss according to the variation of the number of holes in the outer resonate type.

Figure 29 shows the average transmission loss in the low frequency and the entire frequency regions according to the change in the number of holes. The average transmission loss in the low frequency region tends to decrease as the number of holes increases, and the average transmission loss in the entire frequency region tends to increase. Unlike other types, the average transmission loss at low frequencies was higher than the average transmission loss in the entire frequency region. In the low frequency region, the average transmission loss was the highest at 32 dB at 60 holes, and the average transmission loss in the entire frequency region was the highest at 29.6 dB at 150 holes.

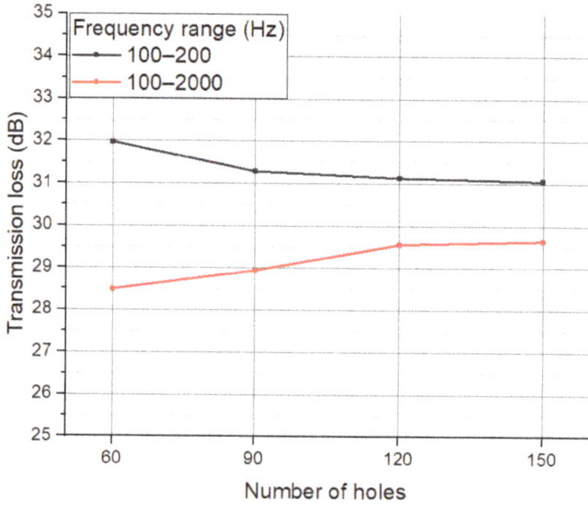

Figure 29. Average transmission loss according to the variation of the number of holes in the outer resonate type.

Figures 30 and 31 show the pressure contour and pressure drop of the scrubber according to the change in the number of holes in the outer resonate type. The number of holes has little effect on the pressure inside the scrubber, and the pressure drop slightly increases as the number of holes increases. The pressure drop is 530.8 Pa at 120 holes and 531.1 Pa at 150 holes, which are the minimum and maximum values.

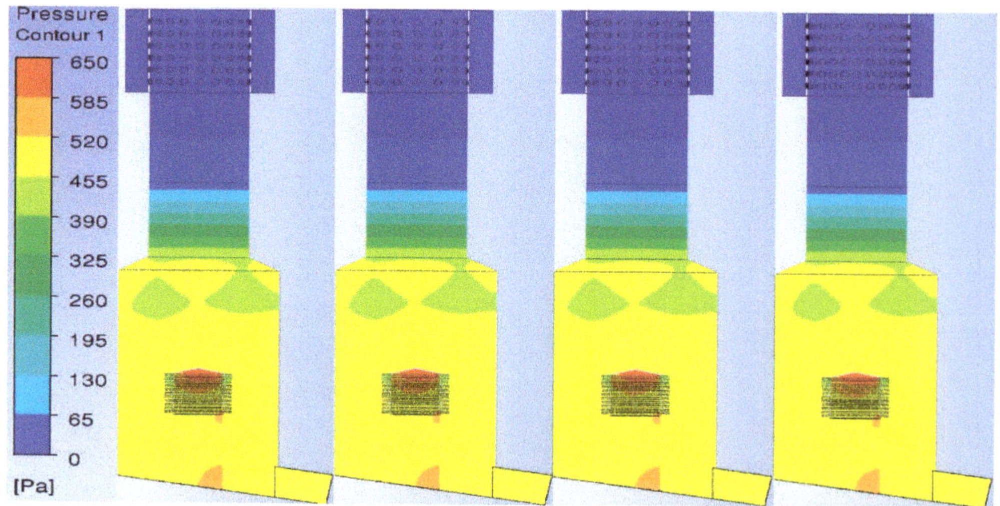

Figure 30. Pressure contour according to the number of holes in the outer resonate type.

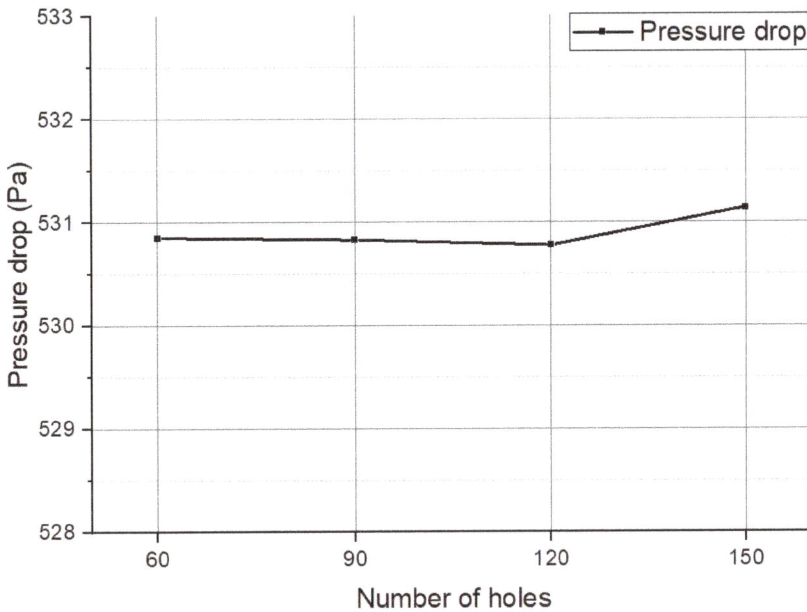

Figure 31. Pressure drop graph according to the number of holes in the outer resonate type.

In the outer resonate type, an increase of the number of holes decreases the average transmission loss in the low frequency region and increases the average transmission loss in the entire frequency region.

3.7. Comparison of Optimal Cases by Type

Figure 32 shows the transmission loss in the entire frequency domain according to the optimal case of each type. Figure 33 shows the pressure drop and the average transmission loss in the low frequency and entire frequency regions. The trend of increasing and decreasing transmission loss was similar for all types, but the outer vane type and outer vane reverse type showed high transmission loss in the 160–500 Hz range. Except for the resonate type, the average transmission loss tends to increase as the pressure drop increases. The pressure drop and average transmission loss were higher in the order of the outer vane type, inner vane type, and resonate type.

Regarding the vane direction in the vane type, both the average transmission loss and pressure drop were higher in the reverse direction than in the forward direction. In the case of the average transmission loss in the entire frequency region, there is a clear difference depending on the direction of the vane, but in the case of average transmission loss at low frequencies, there is no significant difference. In the outer vane reverse type, when the vane length was 225 mm, the pressure drop was 777 Pa, and the average transmission loss in low frequency and entire frequency regions were 43.5 and 54.5 dB, respectively.

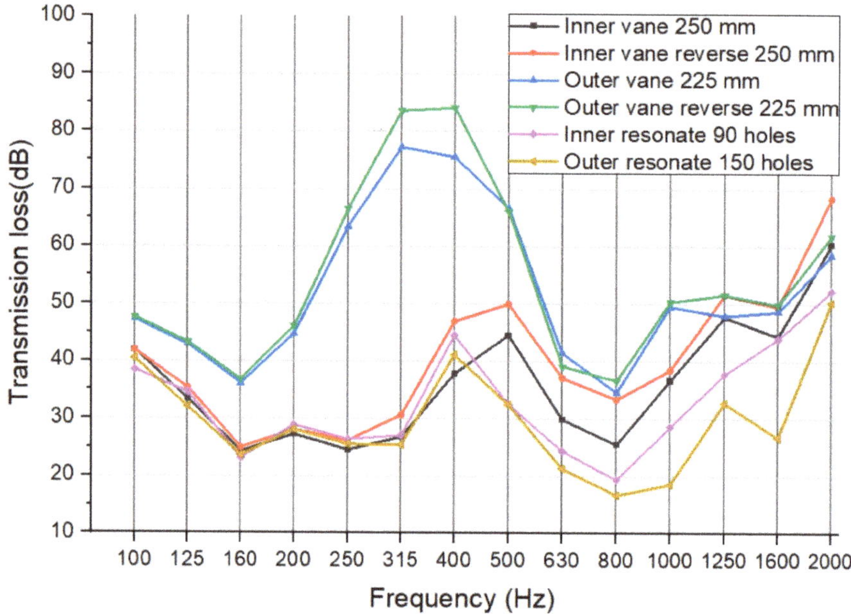

Figure 32. Transmission loss according to the frequency for optimal cases.

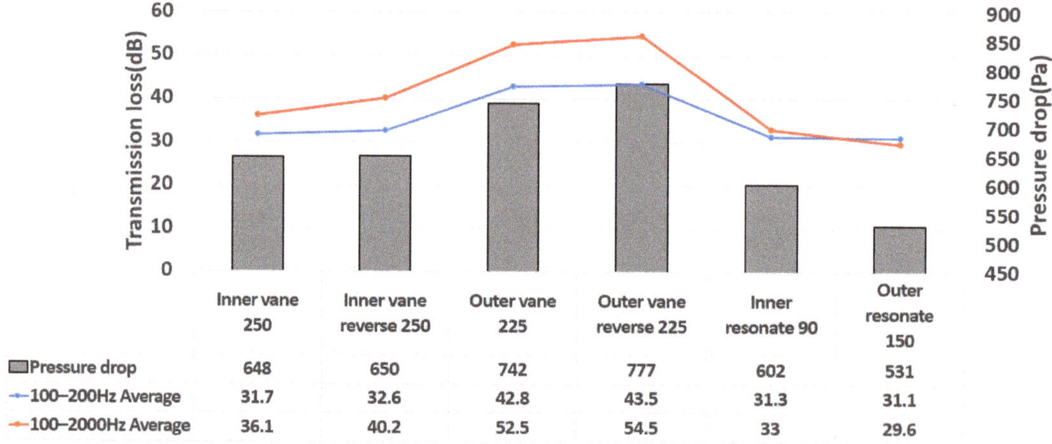

	Inner vane 250	Inner vane reverse 250	Outer vane 225	Outer vane reverse 225	Inner resonate 90	Outer resonate 150
Pressure drop	648	650	742	777	602	531
100–200Hz Average	31.7	32.6	42.8	43.5	31.3	31.1
100–2000Hz Average	36.1	40.2	52.5	54.5	33	29.6

Figure 33. Averaged transmission loss and pressure drop according to the optimal cases for each type.

4. Conclusions

A composite scrubber for ships that simultaneously reduces sulfur oxide and noise was proposed; moreover, the flow characteristics and noise characteristics were analyzed. The results, according to the shape of the silencer in the composite scrubber, are summarized as follows.

(1) In the case of the vane type—as the length of the vane increased, the pressure drop and average transmission loss increased. The pressure drop and average transmission loss were higher when the vane was installed outside rather than inside. They were higher when the vane tilted in the reverse direction rather than the forward direction.

(2) In the case of the inner resonate type—as the number of holes increased, the pressure drop and average transmission loss in the entire frequency region tended to increase, and the average transmission loss in the low frequency region tended to decrease. In the case of the outer resonate type—the pressure drop decreased as the number of holes increased. The optimal performance was shown at 90 holes in the case of the inner resonate type, while the optimal performance was shown at 150 holes in the case of the outer resonate type.

(3) When comparing the optimal cases for each type—the pressure drop and average transmission loss were highest when the vane was installed outside and tilted in the reverse direction (in the case of the vane type).

The condition for maximally reducing noise in the range not exceeding 840 Pa, which is 70% of the allowable engine back pressure, is a vane length of 225 mm in the outer vane reverse type. The pressure drop under this condition was 777 Pa, and the average transmission loss in the entire frequency region was 54.5 dB.

Author Contributions: Conceptualization, M.-r.R. and K.P.; methodology, M.-r.R.; software, M.-r.R.; validation, M.-r.R. and K.P.; formal analysis, M.-r.R.; investigation, M.-r.R.; resources, M.-r.R.; data curation, M.-r.R.; writing-original draft preparation, M.-r.R.; writing-review and editing, K.P.; supervision, K.P.; project administration, M.-r.R. Both authors have read and agreed to the published version of the manuscript.

Funding: This research received no external funding.

Institutional Review Board Statement: Not applicable.

Informed Consent Statement: Not applicable.

Data Availability Statement: Not applicable.

Conflicts of Interest: The authors declare no conflict of interest.

References

1. Lee, S.M.; Kwon, B.M.; Park, K. Flow characteristics according to internal structure of square wet scrubber. *J. Korean Soc. Mar. Eng.* **2019**, *43*, 1–7.
2. Reynolds, K.J. *Exhaust Gas Cleaning Systems Selection Guide, Prepared for Ship Operations Cooperative Program (SOCP)*; The Glosten Associates: Seattle, WA, USA, 2011.
3. Han, B.W.; Kim, H.J.; Kim, Y.J.; Han, K.S. Removal characteristics of gaseous contaminants by a wet scrubber with different packing materials. *J. Korean Soc. Atmos. Environ.* **2007**, *23*, 744–751. [CrossRef]
4. Byeon, S.H.; Lee, B.K.; Mohan, B.R. Removal of ammonia and particulate matter using a modified turbulent wet scrubbing system. *Sep. Purif. Technol.* **2012**, *98*, 221–229. [CrossRef]
5. Lee, B.K.; Mohan, B.R.; Byeon, S.H.; Lim, K.S.; Hong, E.P. Evaluating the performance of a turbulent wet scrubber for scrubbing particulate matter. *J. Air Waste Manag. Assoc.* **2012**, *43*, 499–506. [CrossRef]
6. Son, K.; Lee, J.Y.; Park, K. The effect of spray flow rate, aspect ratio, and filling rate of wet scrubber on smoke reduction. *J. Korean Soc. Mar. Eng.* **2015**, *39*, 217–222.
7. Lee, W.J.; Kim, I.S.; Choi, Y.S.; Choi, J.H. A numerical study on the basic design of scrubber for marine diesel engines. *J. Korean Soc. Mar. Environ. Saf.* **2017**, *23*, 549–557. [CrossRef]
8. Hu, S.Y.; Gao, Y.; Feng, G.; Hu, F.; Liu, C.; Li, J. Experimental study of the dust-removal performance of a wet scrubber. *Int. J. Coal Sci. Technol.* **2021**, *8*, 228–239. [CrossRef]
9. Fang, P.; Cen, C.; Tang, Z.; Zhong, P.; Chen, D.; Chen, Z. Simultaneous removal of SO_2 and NO_X by wet scrubbing using urea solution. *Chem. Eng. J.* **2011**, *168*, 52–59. [CrossRef]
10. Fang, P.; Cen, C.P.; Wang, X.M.; Tang, Z.J.; Tang, Z.X.; Chen, D.S. Simultaneous removal of SO_2, NO and HG^0 by wet scrubbing using urea + $KMnO_4$ solution. *Fuel Process. Technol.* **2013**, *106*, 645–653. [CrossRef]
11. Raghunath, C.V.; Mondal, M.K. Experimental scale multi component absorption of SO_2 and NO by NH_3/NaClO scrubbing. *Chem. Eng. J.* **2017**, *314*, 537–547. [CrossRef]
12. Yang, S.; Pan, X.; Han, Z.; Zhao, D.; Liu, B.; Zheng, D. Removal of NO_X and SO_2 from simulated ship emissions using wet scrubbing based on seawater electrolysis technology. *Chem. Eng. J.* **2018**, *331*, 8–15. [CrossRef]
13. Meikpa, B.C.; Kundu, G.; Biswas, M.N. Modeling of a novel multi-stage bubble column scrubber for flue gas desulfurization. *Chem. Eng. J.* **2002**, *86*, 331–342. [CrossRef]
14. D'Addio, L.; Carotenuto, C.; Balachandran, W.; Lancia, A.; Di Natale, F. Experimental analysis on the capture of submicron particles (PM0.5) by wet electrostatic scrubbing. *Chem. Eng. Sci.* **2014**, *106*, 222–230. [CrossRef]

15. Park, H.W.; Choi, S.S.; Park, D.W. Simultaneous treatment of NO and SO_2 with aqueous $NaClO_2$ solution in a wet scrubber combined with a plasma electrostatic precipitator. *J. Hazard. Mater.* **2015**, *285*, 117–126. [CrossRef]
16. Lim, J.Y.; Kim, H.S.; Bang, S.Y.; Bum, B.S.; Kim, J.H. Effects on the removal of sulfur compounds by microbubble supply to circulation tank in a scrubber. *J. Odor Indoor Environ.* **2019**, *18*, 10–17. [CrossRef]
17. Sun, W.; Shao, Y.; Zhao, L.; Wang, Q. Co-removal of CO_2 and particulate matter from industrial flue gas by connecting an ammonia scrubber and a granular bed filter. *J. Clean. Prod.* **2020**, *257*, 120511. [CrossRef]
18. Xie, R.; Ji, J.; Guo, K.; Lei, D.; Fan, Q.; Leung, D.Y.C.; Huang, H. Wet scrubber coupled with UV/PMS process for efficient removal of gaseous VOCs: Roles of sulfate and hydroxyl radicals. *Chem. Eng. J.* **2019**, *356*, 632–640. [CrossRef]
19. Lee, S.H.; Kim, N.S.; Jin, B.M.; You, J.C.; Do, C.S. Sound insulation case of diesel engine low frequency exhaust noise problem. In Proceedings of the Annual Autumn Meeting, Mokpo, Korea, 3–4 November 2011; pp. 634–637.
20. Hwang, W.G.; Lee, Y.Y.; Oh, J.E.; Kim, K.S.; Song, Y.H. A Study on the Transmission Loss and Back pressure of Muffler Elements. *J. Korean Soc. Precis. Eng.* **2002**, *19*, 147–153.
21. Kwon, J.; Jeong, W.B.; Hong, C.S. Effects of Fluid Velocity on Acoustic Transmission Loss of Simple Expansion Chamber. *Trans. Korean Soc. Noise Vib. Eng.* **2012**, *22*, 994–1002. [CrossRef]
22. Kim, Y.W.; Jeong, W.B. Numerical Analysis of Noise Reduction and Back-pressure for a Simple Expansion Chamber with a Partition. *Trans. Korean Soc. Noise Vib. Eng.* **2014**, *24*, 883–889. [CrossRef]
23. Yi, C.S.; Lee, T.E.; Lee, C.W. Numerical analysis of the Internal Flow of 8kW Grade Diesel Generator Muffler. *J. Korean Soc. Manuf. Process. Eng.* **2018**, *17*, 45–50. [CrossRef]
24. Secgin, E.; Arslan, H.; Birgören, B. A statistical design optimization study of a multi-chamber reactive type silencer using simplex centroid mixture design. *J. Low Freq. Noise Vib. Act. Control* **2020**, *40*, 623–638. [CrossRef]
25. Nursal, R.S.; Hashim, A.H.; Nordin, N.I.; Hamid, M.A.A.; Danuri, M.R. CFD Analysis of the Effects of Exhaust Backpressure Generated by Four-Stroke Marine Diesel Generator After Modification of Silencer and Exhaust Flow Design. *ARPN J. Eng. Appl. Sci.* **2017**, *12*, 1271–1280.
26. Kakade, S.K.; Sayyad, F.B. Optimization of Exhaust Silencer for Weight and Size by Using Noise Simulation for Acoustic Performance. *J. Inf. Comput. Sci.* **2017**, *10*, 49–60.
27. Kim, S.A. Silencer structure for exhaust noise attenuation performance in the high frequency range. *J. Korean Soc. Fish. Ocean Technol.* **2018**, *54*, 255–261. [CrossRef]
28. ANSYS CFX-Theory Guide V14.0. Available online: https://www.researchgate.net/publication/321628826_Study_on_Design_of_Pressure_Chamber_in_a_Linear-Jet_Type_Air_Curtain_System_for_Prevention_of_Smoke_Spread/fulltext/5a29ec470f7e9b63e5353c2d/Study-on-Design-of-Pressure-Chamber-in-a-Linear-Jet-Type-Air-Curtain-System-for-Prevention-of-Smoke-Spread.pdf (accessed on 1 September 2021).
29. *ANSYS AACTx_R170_L*. V17.0. Available online: https://www.ansys.com/news-center/press-releases/01-27-16-ansys-unveils-release-17-0 (accessed on 1 September 2021).
30. Strigle, R.F., Jr. Random and Structured Packings. In *Packed Tower Design and Applications*, 2nd ed.; Gulf Publishing Company: Houston, TX, USA, 1987.
31. Kang, D.J.; Lee, W.S.; Lee, J.W.; Hong, J.K.; Jo, Y.H.; Lee, S.J. *Survey of the Soundproof Materials Effect—On Absorptive Materials*. In *Survey of the Soundproof Materials Effect*; Air Quality Research Department, National Institute of Environmental Research: Hwangyong-ro, Korea, 2004; MONO1200503777. Available online: http://dl.nanet.go.kr/law/SearchDetailView.do?cn=MONO1200503777 (accessed on 6 September 2021).

Journal of
*Marine Science
and Engineering*

Article

Increasing Mechanical Properties of 3D Printed Samples by Direct Metal Laser Sintering Using Heat Treatment Process

Jozef Živčák [1], Ema Nováková-Marcinčínová [1,*], Ľudmila Nováková-Marcinčínová [2], Tomáš Balint [1] and Michal Puškár [1]

[1] Faculty of Mechanical Engineering, Technical University of Košice, Letná 9, 042 00 Košice, Slovakia; jozef.zivcak@tuke.sk (J.Ž.); tomas.balint@tuke.sk (T.B.); michal.puskar@tuke.sk (M.P.)
[2] Faculty of Manufacturing Technologies with a Seat in Prešov, Technical University of Kosice, Bayerova 1, 080 01 Prešov, Slovakia; ludmila.novakovamarcincinova@tuke.sk
* Correspondence: ema.novakova-marcincinova@tuke.sk; Tel.: +421-055-6023543

Abstract: The paper deals with the evaluation of mechanical properties of 3D-printed samples based on high-strength steel powder system maraging steel using direct metal laser sintering (DMLS), which is currently being put into technical practice. The novelty of this article is that it analyzes mechanical properties of samples both printed and age hardened as well as examining the fracture surfaces. When comparing the manufacturer's range with our recorded values, samples from Set 1 demonstrated strength ranging from 1110 to ultimate 1140 MPa. Samples from Set 2 showed tensile strength values that were just below average. Our recorded range was from 1920 to ultimate 2000 MPa while the manufacturer reported a range from 1950 to 2150 MPa. The tensile strength was in the range from 841 to ultimate 852 MPa in Set 1, and from 1110 to ultimate 1130 MPa in Set 2.

Keywords: mechanical properties; tensile strength; laser sintering; fracture; additive technology; rapid prototyping

Citation: Živčák, J.; Nováková-Marcinčínová, E.; Nováková-Marcinčínová, Ľ.; Balint, T.; Puškár, M. Increasing Mechanical Properties of 3D Printed Samples by Direct Metal Laser Sintering Using Heat Treatment Process. *J. Mar. Sci. Eng.* **2021**, *9*, 821. https://doi.org/10.3390/jmse9080821

Academic Editor: María Isabel Lamas Galdo

Received: 5 July 2021
Accepted: 27 July 2021
Published: 30 July 2021

Publisher's Note: MDPI stays neutral with regard to jurisdictional claims in published maps and institutional affiliations.

1. Introduction

The idea of transforming the innovative production processes from a prototype into serial production has recently become increasingly popular for many manufacturers. EOS GmbH (Electro-Optical Systems) is the global technology and quality leader for high-end solutions in the field of additive manufacturing (AM) and a global player in the field of direct metal laser sintering (DMLS). Direct metal laser sintering was developed jointly by Rapid Product Innovations (RPI) and EOS GmbH, (Electro-Optical Systems) starting in 1994 as the first commercial rapid prototyping method to produce metal parts in a single process. In addition to functional prototypes, components made by DMLS are often used for rapid tool making, medical implants, aerospace parts, and components for high-temperature environments. Additive manufacturing allows more cost-effective production for single part production, rather than batch production.

In DMLS technology, metal powder of 20 microns, free of binder or fluxing agent, is completely melted by scanning using a high-performance laser beam to form a part with original material properties. Elimination of the polymeric binder avoids the burn-off and infiltration steps and produces 95% dense steel compared to about 70% density with selective laser sintering (SLS) [1–10].

An additional benefit of the DMLS process compared to SLS is higher detail resolution due to the use of thinner layers, enabled by a smaller powder diameter. This capability allows for more intricate part shapes [1–8,11]. Material options that are currently offered include alloy steel, stainless steel, tool steel, aluminum, bronze, cobalt–chrome, and titanium [3–11]. The complexity of these parts has almost no effect on production time and costs with this technology. Complex lightweight constructions can often reduce weight and save material costs. The EOS process has proven itself in practice and can benefit from faster and more cost-effective production: 43% shorter cooling time, 31% shorter cycle time, 43% shorter injection molding time, and functionality [4,12].

Several authors have observed mechanical properties of powder additive technologies. Recently, heat treatment methods have been used to improve these properties, where hardness and strength values have increased by more than twice [13–15], and several reports have been published on investigating the characteristics of Ti6Al4V materials produced by AM layering. The work of Brandl et al. evaluates the fatigue strength of Ti6Al4V material prepared by the laser melting process [16]. The force deformation characteristics of Ti6Al4V produced by DMLS and SLS were measured according to Facchini et al. [17]. Fatigue stress of Ti6Al4V samples produced by SLM was investigated by Van Hooreweder [18]. Liu et al. presented fatigue in Ti6Al4V samples produced by SLM, attributing weak fatigue behavior to defects [19]. Comparison of fatigue strength of samples prepared from Ti6Al4V by DMLS using the EOSINT M270 machine compared to the e-beam system was presented by Chan et al. [20]. Rafi et al. measured tensile and fatigue properties of both Ti6Al4V samples manufactured by EOS M270. Published reports on mechanical properties of high-strength steels manufactured by AM techniques are much more limited. Rafi et al. [21] and Edwards et al. [22] included DMLS materials and stainless steel in their research. The aim of the work was to evaluate mechanical properties of the newly used system of high-strength steel, to analyze the fracture surface after a static tensile test as well as to point out the redistribution of particles influenced by diversity of the chemical composition of the material used. The tests dealt with research of mechanical properties of the 3D-printed samples based on high-strength steel powder system Ti6Al4V maraging steel using direct metal laser sintering (DMLS), which is currently being put into technical practice. This research was developed based on cooperation with 1.Prešovská nástrojáreň as necessary for technical practice in order to verify and improve the properties of the mentioned material for the possibilities of better production of prototypes as well as products using the DMLS method.

2. Materials and Methods

2.1. Research Materials

The powder material from which the samples were prepared was MS1 Maraging Steel, a trademark of EOS GmBH Company. EOS Maraging Steel MS1 is a steel powder that has been specially optimized for processing on the 200W EOSINT M 280. The material composition of MS1 is listed in Table 1.

Table 1. Material composition.

Ni %	Co %	Mo %	Ti %	Al %	Cr, Cu %	C %	Mn, Si %	P, S %	Fe %
17–19	8.5–9.5	4.5–5.2	0.6–0.8	0.05–0.15	Each ≤ 0.5	≤0.03	Each ≤ 0.1	≤0.01	balance

Parts made of EOS Maraging Steel MS1 have chemical composition corresponding to the US classification of 18% Ni Maraging 300, European Norm 1.2709 and German X3NiCoMoTi 18-9-5. This type of steel is characterized by very good mechanical properties and is easily heat treatable using a simple hardening process to achieve excellent hardness and strength. Parts manufactured from EOS Maraging Steel MS1 are easily machined after the building process and can be easily hardened to more than 50 HRC at 490 °C (914 °F) for 6 h. In both as-built and age-hardened states, the parts can be further machined using chipping, electrical discharge machining, welding, sanding, polishing and coating, if necessary. As a result of the layer-forming method, these parts have some anisotropy, which can be reduced or eliminated by suitable heat treatment. Its mechanical properties with EOS production are listed in Tables 2 and 3.

Table 2. Mechanical properties of Ti6Al4V: (**a**) as built, only printed, (**b**) printed and age-hardened [9].

Density ρ, g/cm^3		Tensile Strength MPa		Yield Strength Rp 0.2 %		Elongation at Break % *		Modulus of Elasticity GPa *		Hardness HRC **
direction		XY	Z	XY	Z	XY	Z	XY	Z	
8.0–8.1	(a)	1100 ± 100	1100 ± 100	1050 ± 100	1000 ± 150	10 ± 4	-	160 ± 25	150 ± 20	33–37
	(b)	2050 ± 100		1190 ± 100		4 ± 2				50–56

* Tensile testing according to ISO 6892-1:2009. (B) Annex D, proportional test pieces, diameter of the neck area 5 mm (0.2 inch), original gauge length 25 mm (1 inch). ** Rockwell C (HRC) hardness measurement according to EN ISO 6508-1 on polished surface. Note that measured hardness can vary significantly depending on how the specimen has been prepared.

Table 3. Thermal properties of parts—Ti6Al4V [9].

	As Built	After Age Hardening ***
Thermal conductivity	15 ± 0.8 W/m °C	20 ± 1 W/m °C
Specific heat capacity	450 ± 20 J/kg °C	450 ± 20 J/kg °C
Maximum operating temperature		approx. 400 °C

*** Ageing temperature 490 °C (914 °F), 6 h, air cooling.

The surface roughness of the material Ti6Al4V is approximately Ra = 5 μm and Rz = 28 μm with the MS1 performance method (40 μm) at MS1 surface (20 μm) Ra = 4 μm and Rz = 28 μm. Due to the layerwise building, the surface structure depends on the orientation of the surface; for example, sloping and curved surfaces are shown as a stair-step effect. These values also depend on the measurement method used. The above values indicate the difference for the horizontal orientation of the sample (upward—XY) or vertical (Z) orientation. The method used was MS1 surface and the layering was in the XY plane. Samples were prepared according to the ASTM standard for powder metallurgy technology research. Subsequently, the fracture surface of the samples was analyzed using a scanning electron microscope.

2.2. Methods of Research

The methodology of investigating the properties of printed materials determines and evaluates the type of high-strength steel described in the previous chapter. Tested samples were obtained by direct metal laser sintering with 3D printing in additive technologies. Two series were prepared, each with 20 samples: only printed—as-built (Set 1), and printed and age-hardened (Set 2). We followed the EOS printer manufacturer's recommendations for the hardening process (Figure 1). Powder was also produced by the same manufacturer. The hardening process was carried out in a hardening furnace at 490 °C for 6 h; cooling was carried out in an operating room with room temperature within one day.

Strength and elongation tests were performed on samples according to ASTM standard [11] at the Technical University of Košice.

Samples were made in tool shop 1, Presovska Nastrojaren LTD, Prešov, Slovak Republic. Dimensions and shapes of the test samples are shown in Figure 2 and Table 4.

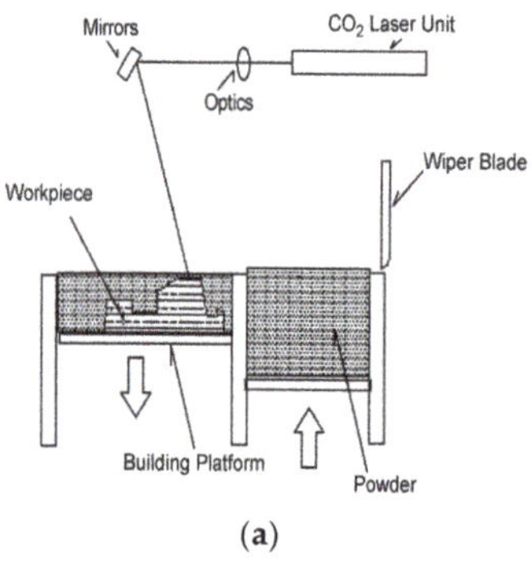

(**a**)

(**b**)

Figure 1. A schematic diagram [15] of the EOS Machine (**a**). The final sintering process in the production of samples on a DMLS 3D printer (**b**).

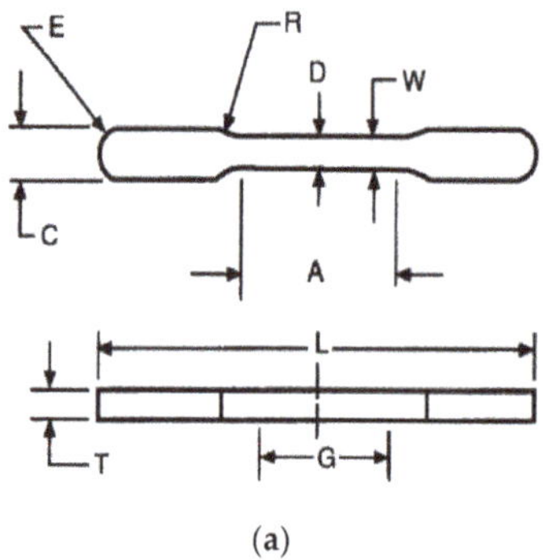

(**a**)

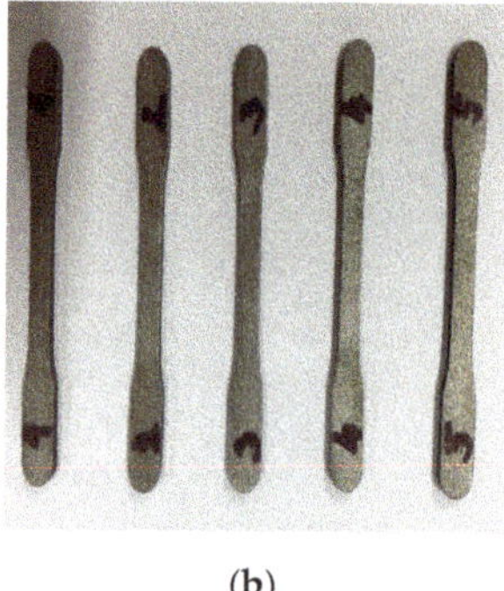

(**b**)

Figure 2. Flat unmachined tension test specimen [11] (**a**) and real view of samples (**b**) after sintering process.

Table 4. Dimension of tension test specimen by ASTM [10].

	Dimension [inch]	**[mm]**
G—Gage length	1.000 ± 0.003	25.40 ± 0.08
L—Overall length	3.53	89.7
C—Width of grip section	0.34	8.6
E—End radius	C/2	C/2
W—Width of reduced section	0.235	5.97
D—Width at center	0.225	5.72
A—Length of reduced section	1.25	31.8
R—Radius of fillet	1.00	25.4
T—Thickness	0.140 to 0.250	3.56 to 6.35

Study of the fracture surface morphology was performed on the scanning electron microscope TESCAN MIRA 3 (IMR SAS, Košice, Slovakia).

3. Results and Discussion

Based on the static tensile test, mechanical properties of the two series of samples of five pieces each, which are summarized in Table 5, were examined. According to ISO 6892-1 [13], the Tinius Olsen hydraulic machine was used for the tensile test. Figure 3 illustrates graphs of a portion of the tests of (a) as-built samples (Set 1) and (b) printed and age-hardened samples (Set 2). Samples from (b) Set 2 showed a significant increase in strength values.

Table 5. Summarized mechanical properties from tensile tests.

Number of Samples	Yield Tensile Strength, MPa	Ultimate Tensile Strength, MPa	Elongation at Break %
Set 1			
Samples 1–5	841–852	1110–1140	11.1–14.3
Set 2			
Samples 1–5	1110–1130	1920–2000	2.35–3.27

Force sensor: 300 kN, speed of deformation: $0.0001\ \mathrm{s}^{-1}$.

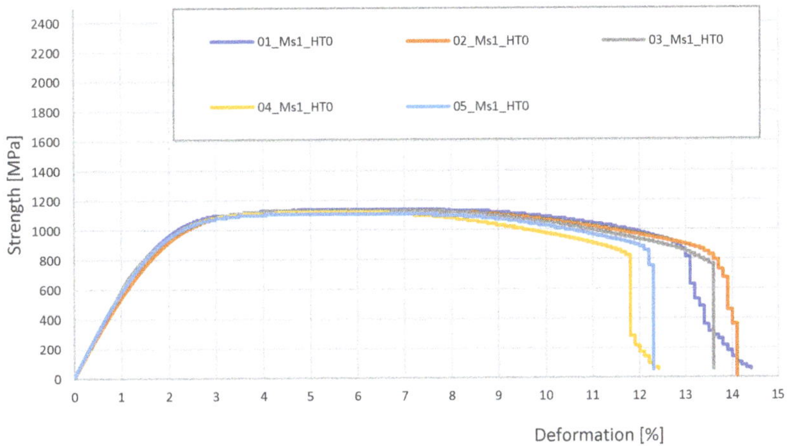

(**a**) Samples of Set 1, as-built, samples 01–05 Ms1 HTD.

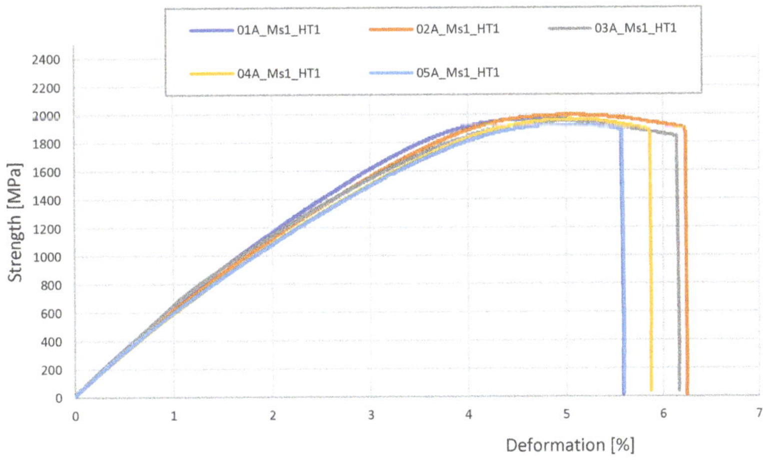

(**b**) Samples of Set 2, age-hardened, samples 01–05 Ms1 HT1.

Figure 3. Strength tests report.

If we compare the range given by the EOS printer manufacturer with our recorded values, the samples from Set 1 demonstrated ultimate strength ranging from 1110 to 1140 MPa.With the samples from the Set 2, the tensile strength values were only slightly below the average. Our recorded range was from 1920 to 2000 MPa, as opposed to the recommended range given by EOS of 1950–2150 MPa.

The tensile strength range was from 841 to 852 MPa in Set 1, and 1110–1130 MPa in Set 2. We can state that elongation values were within this range. The only exception was the samples from the Set 1, where the values slightly exceeded 14%. Although the hardening process is ideal for increasing hardness, as it is presented in [13,14], where hardness increased from 33 HRC to 54 HRC, elongation and tensile strength values are significantly reduced.

Fracture surface morphology was also evaluated after static tensile testing of the samples. They were observed by a scanning electron microscope and surfaces are shown in the Figures 4–6. The cross section showed a neck variation that is only visible on the printed sample (Set 1). The samples that were printed and then hardened (Set 2) demonstrated a less significant area compared to the Set 1 samples. The samples from Set 2 showed more fragile sites during surface morphology examination while the observed refractive surface of the samples from Set 1 were more plastic. In both cases, the different morphology of the whole surface was observed.

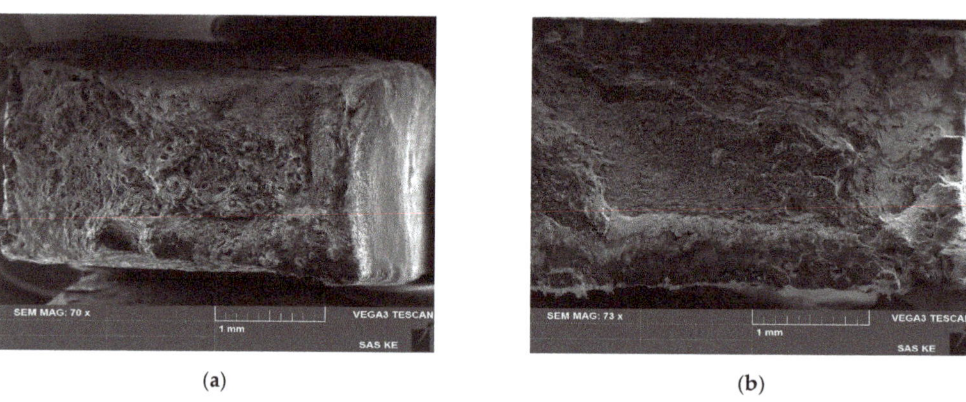

(**a**)　　　　　　　　　　　　　　　　　(**b**)

Figure 4. Surface of fracture samples: (**a**) Set 1, (**b**) Set 2 (×70). Scale bar: 1 mm.

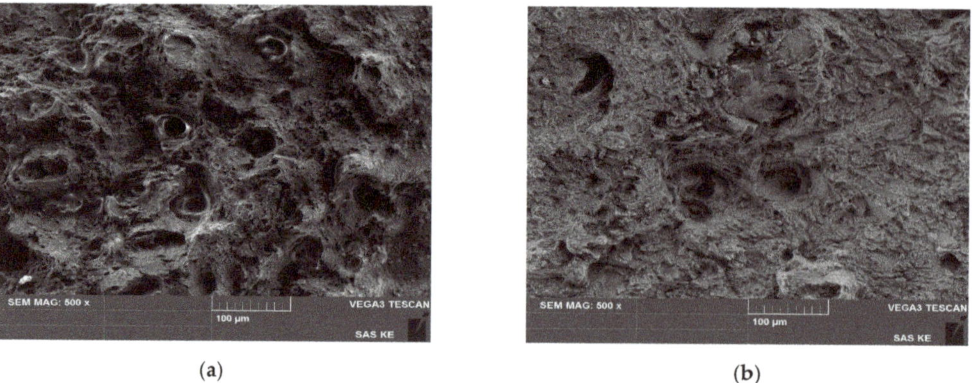

(**a**)　　　　　　　　　　　　　　　　　(**b**)

Figure 5. Surface of fracture samples: (**a**) Set 1, (**b**) Set 2 (×500). Scale bar: 100 μm.

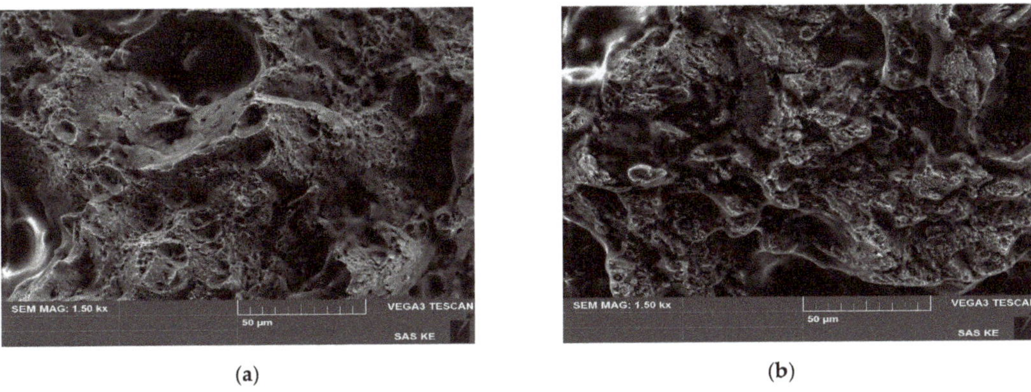

(a)   **(b)**

Figure 6. Surface of fracture samples: (**a**) Set 1, (**b**) Set 2 (×1500). Scale bar: 50 μm.

Table 6 presents the measured results of the chemical composition of the material Ti6Al4V of both series of samples, which were examined by EDS analysis on the evaluated fracture surfaces.

Table 6. Measured values of EDS analysis of the chemical composition of the material Ti6Al4V.

Spectrum: 1x

El	AN	Series	unn. C	norm. C	Atom. C
			[wt.%]	[wt.%]	[at.%]
O	8	K-series	10.15	13.23	34.63
S	16	K-series	1.54	2.01	2.63
Ti	22	K-series	0.61	0.80	0.70
Fe	26	K-series	46.02	59.98	44.97
Co	27	K-series	7.32	9.54	6.78
Ni	28	K-series	11.08	14.44	10.30
		Total:	76.73	100.00	100.00

Spectrum: 2x

El	AN	Series	unn. C	norm. C	Atom. C
			[wt.%]	[wt.%]	[at.%]
O	8	K-series	11.00	13.71	36.43
Ti	22	K-series	0.62	0.78	0.69
Fe	26	K-series	46.67	58.17	44.28
Co	27	K-series	7.4	9.22	6.65
Ni	28	K-series	11.18	13.94	10.09
Mo	42	L-series	3.36	4.19	1.86
		Total:	76.73	100.00	100.00

Spectrum: 3x

El	AN	Series	unn. C	norm. C	Atom. C
			[wt.%]	[wt.%]	[at.%]
O	8	K-series	7.54	7.61	22.35
S	16	K-series	1.25	1.26	1.84
Ti	22	K-series	1.28	1.29	1.27

Table 6. *Cont.*

El	AN	Series	unn. C	norm. C	Atom. C
Fe	26	K-series	64.82	65.42	55.03
Co	27	K-series	11.08	11.18	8.91
Ni	28	K-series	13.11	13.24	10.59
		Total:	99.09	100.00	100.00

Spectrum: 4x

El	AN	Series	unn. C	norm. C	Atom. C
			[wt.%]	[wt.%]	[at.%]
O	8	K-series	7.77	7.80	23.19
Ti	22	K-series	1.25	1.25	1.24
Fe	26	K-series	64.24	64.43	54.91
Co	27	K-series	10.92	10.96	8.85
Ni	28	K-series	12.92	12.96	10.51
Mo	42	L-series	2.60	2.61	1.29
		Total:	99.71	100.00	100.00

Spectrum: 5x

El	AN	Series	unn. C	norm. C	Atom. C
			[wt.%]	[wt.%]	[at.%]
O	8	K-series	7.21	8.60	25.31
Fe	26	K-series	52.62	62.72	52.89
Co	27	K-series	8.25	9.84	7.86
Ni	28	K-series	12.66	15.09	12.11
Mo	42	L-series	3.14	3.75	1.84
		Total:	83.9	100.00	100.00

Spectrum: 6x

El	AN	Series	unn. C	norm. C	Atom. C
			[wt.%]	[wt.%]	[at.%]
O	8	K-series	8.46	9.89	28.27
Ti	22	K-series	0.66	0.77	0.73
Fe	26	K-series	52.88	61.81	50.63
Co	27	K-series	8.16	9.54	7.41
Ni	28	K-series	12.38	14.47	11.28
Mo	42	L-series	3.01	3.52	1.68
		Total:	85.54	100.00	100.00

We have recorded the summary range mass percentage in the following range: Ti 0.69–1.27, Co 6.65–8.91, Ni 10.09–12.11, Mo 1.29–1.86. The observed Ni was below the manufacturer's reported average. The manufacturer reports 19% by weight. We have recorded that it is not exceeding 12.11% by weight. In contrast, Ti exceeded its range; it was above average. The recorded value was up to 1.27% by weight M4, while the manufacturer gives a range of up to 0.8% by weight. The difference found is fundamental for the mechanical properties of products made of this material. These different limits may cause different values of the reported mechanical properties specified by the manufacturer.

4. Conclusions

The following important research results have been identified:

- Although the hardening process is ideal for increasing hardness from 33 to ultimate 54 HRC [10], elongation and tensile strength values are significantly reduced. When comparing the manufacturer's range with our recorded values, samples from Set 1 demonstrated strength ranging from 1110 to ultimate 1140 MPa. Samples from Set 2 showed tensile strength values that were just below average. Our recorded range was from 1920 to ultimate 2000 MPa while the manufacturer reported a range from 1950 to 2150 MPa. The tensile strength was in the range from 841 to ultimate 852 MPa in Set 1, and from 1110 to ultimate 1130 MPa in Set 2. The elongation values can be found to be within the range of 14% for the Set 1 samples, or slightly over 14%.
- Different limits measured in EDS chemical composition analysis may cause different values of reported mechanical properties. The observed Ni was below the manufacturer's average. We noticed that it did not exceed 12.11% by weight. Conversely, Ti exceeded its range, it was above average, recording a value of up to 1.27% by weight while the manufacturer's specified range is up to 0.8% by weight.
- The cross section of the samples showed a difference in neck that was only visible on the printed sample (Set 1). Samples that were printed and further hardened (Set 2) had a less significant area compared to the Set 1 samples. The samples from Set 2 showed more fragile areas. In contrast, the observed fracture surface of the samples from Set 1 was more plastic in both, with differences of the whole morphology.

However, despite experimentally measured values showing lower output values than the results of measurements from the material manufacturer for EOS 3D printers, it is known that these findings do not seriously affect quality of production. The strength of Ti6Al4V products can be further increased by furnace curing, like samples from the Set 2 series in our experiment. This allows unique production of form complexity and even stronger metal components than production using conformal cooling or repair of broken parts using metal powder.

Author Contributions: Conceptualization, J.Ž., E.N.-M. and Ľ.N.-M.; methodology, J.Ž. and E.N.-M.; investigation, M.P., Ľ.N.-M. and T.B.; data analysis, E.N.-M. and Ľ.N.-M.; writing—original draft preparation, E.N.-M. and Ľ.N.-M. and T.B.; writing—review and editing, M.P. All authors have read and agreed to the published version of the manuscript.

Funding: This work was supported by the Slovak Research and Development Agency under the Contract no. APVV-19-0290. The article was written in the framework of Grant Projects KEGA 041TUKE-4/2019, KEGA 006TUKE-4/2020 and the project Stimuls for research and development no. 2018/14432:1-26CO supported by the Ministry of Education of Science of Research.

Institutional Review Board Statement: Not applicable.

Informed Consent Statement: Not applicable.

Data Availability Statement: Not applicable.

Conflicts of Interest: The authors declare no conflict of interest.

References

1. Mancanares, C.G.; Zancul, E.D.; da Silva, J.C.; Miguel, P.A.C. Additive manufacturing process selection based on parts' selection criteria. *Int. J. Adv. Manuf. Technol.* **2015**, *80*, 1007–1014. [CrossRef]
2. Mower, T.M.; Long, M.J. Mechanical behavior of additive manufactured, powder-bed laser-fused materials. *Mat. Sci. Eng.* **2015**, *651*, 198–213. [CrossRef]
3. Nováková-Marcinčínová, L.; Novák-Marcinčín, J. Application of rapid prototyping fused deposition modelling materials. In Proceedings of the 23rd International DAAAM Symposium, Zadar, Croatia, 24–27 October 2012; pp. 57–60.
4. Nováková-Marcinčínová, L. Application of fused deposition modeling technology in 3D printing rapid prototyping area. *Manuf. Ind. Eng.* **2012**, *11*, 35–37.
5. Yasa, E.; Kruth, J.-P. Microstructural investigation of selective laser melting 316L stainless steel parts exposed to laser re-melting. *Procedia Eng.* **2011**, *19*, 389–395. [CrossRef]

6.   Dzian, A.; Zivcak, J.; Penciak, R.; Hudak, R. Implantation of a 3D-printed titanium sternum in a patient with a sternal tumor. *World J. Surg. Oncol.* **2018**, *16*, 7. [CrossRef] [PubMed]

7.   Findrik Balogova, A.; Hudak, R.; Toth, T.; Schnitzer, M.; Ferenc, J.; Bakos, D.; Zivcak, J. Determination of geometrical and viscoelastic properties of PLAPHB samples made by additive manufacturing for urethral substitution. *J. Biotechnol. Rev. Mol. Biotechnol.* **2018**, *284*, 123–130. [CrossRef]

8.   Guzova, A.; Draganovská, D.; Ižarikova, G.; Živcak, J.; Hudak, R.; Brezinova, J.; Moro, R. The Effect of Position of Materials on a Build Platform on the Hardness, Roughness, and Corrosion Resistance of Ti6Al4V Produced by DMLS Technology. *Metals* **2019**, *10*, 1055. [CrossRef]

9.   Guzanová, A.; Ižaríková, G.; Brezinová, J.; Živčák, J.; Draganovská, D.; Hudák, R. Influence of Build orientation, heat treatment, and laser power on the hardness OF Ti6al4v manufactured using The dmls Process. *Metals* **2017**, *7*, 318. [CrossRef]

10. Gál, P.; Toporcer, T.; Vidinský, B.; Hudák, R.; Živčák, J.; Sabo, J. Simple interrupted PERCUTANEOUS Suture versus Intradermal Running Suture for WOUND tensile strength measurement in Rats: A technical note. *Eur. Surg. Res.* **2009**, *43*, 61–65. [PubMed]

11. Hudák, R.; Šarik, M.; Dadej, R.; Živčák, J.; Harachová, D. Material and thermal analysis of laser sintered products. *Acta Mech. et Autom.* **2013**, *7*, 15–19.

12. Sabol, F.; Vasilenko, T.; Novotný, M.; Tomori, Z.; Bobrov, N.; Živčák, J.; Hudák, R.; Gál, P. Intradermal running Suture versus 3M™ vetbond™ Tissue adhesive for Wound closure in RODENTS: A BIOMECHANICAL and histological study. *Eur. Surg. Res.* **2010**, *45*, 321–326. [CrossRef] [PubMed]

13. Toth, T.; Hudak, R.; Zivcak, J. Dimensional verification and quality control of implants produced by additive manufacturing. *Qual. Innov. Prosper.* **2015**, *19*, 9–21. [CrossRef]

14. Tóth, T.; Živčák, J. A comparison of the outputs of 3d scanners. *Procedia Eng.* **2014**, *69*, 393–401. [CrossRef]

15. Živčák, J.; Šarik, M.; Hudák, R. FEA simulation of thermal processes during the Direct metal laser sintering of TI64 titanium powder. *Measurement* **2016**, *94*, 893–901. [CrossRef]

16. Brandl, E.; Leyens, C.; Palm, F.; Schoberth, A.; Onteniente, P. Wire instead of powder? Properties of additive manufactured Ti–6Al–4V for aerospace applications. In Proceedings of the Euro-uRapid International Conference on Rapid Prototyping, Rapid Tooling & Rapid Manufacturing, Berlin, Germany, 23–24 September 2008.

17. Facchini, L.; Magalini, E.; Robotti, P.; Molinari, A.; Höges, S.; Wissenbach, K. Ductility of a Ti–6Al–4V alloy produced by selective laser melting of pre-alloyed powders. *Rapid Prototyp. J.* **2010**, *16*, 450–459. [CrossRef]

18. Van Hooreweder, B.; Boonen, R.; Moens, D.; Kruth, J.-P.; Sas, P. On the determination of fatigue properties of Ti6Al4V produced by selective laser melting. In Proceedings of the 53rd AIAA/ASME Structures, Structural Dynamics and Materials Conference, Honolulu, HI, USA, 23–26 April 2012.

19. Liu, Q.; Elambasseril, J.; Sun, S.; Leary, M.; Brandt, M.; Sharp, P.K. The effect of manufacturing defects on the fatigue behavior of Ti–6Al–4V specimens fabricated using selective laser melting. *Adv. Mater. Res.* **2014**, *891*, 1519–1524. [CrossRef]

20. Chan, K.S.; Koike, M.; Mason, R.L.; Okabe, T. Fatigue life of titanium alloys fabricated by additive layer manufacturing techniques for dental implants. *Metall. Mater. Trans. A* **2013**, *44*, 1010–1022. [CrossRef]

21. Rafi, H.K.; Starr, T.L.; Stucker, B.E. A comparison of the tensile, fatigue and fracture behavior of Ti–6Al–4V and 15–5 PH stainless steel parts made by selective laser melting. *Int. J. Manuf. Technol.* **2013**, *69*, 1299–1309. [CrossRef]

22. Edwards, P.; Ramulu, M. Fatigue performance of selective laser melted Ti–6Al–4V. *Mater. Sci. Eng. A* **2014**, *598*, 327–337. [CrossRef]

Journal of
*Marine Science
and Engineering*

Article

Acoustic Method for Estimation of Marine Low-Speed Engine Turbocharger Parameters

Roman Varbanets [1], Oleksij Fomin [2], Václav Píštěk [3], Valentyn Klymenko [1], Dmytro Minchev [4], Alexander Khrulev [5], Vitalii Zalozh [6] and Pavel Kučera [3],*

[1]   Marine Engineering Department, Odessa National Maritime University, 34 Mechnikov Str., 65029 Odessa, Ukraine; seute.head@onmu.odessa.ua (R.V.); seute.klymenko@onmu.odessa.ua (V.K.)
[2]   Department of Cars and Carriage Facilities, State University of Infrastructure and Technologies, Kyrylivska Str. 9, 04071 Kyiv, Ukraine; fomin_ov@gsuite.duit.edu.ua
[3]   Institute of Automotive Engineering, Brno University of Technology, Technická 2896/2, 616 69 Brno, Czech Republic; pistek.v@fme.vutbr.cz
[4]   Department of Internal Combustion Engines, Plants and Technical Maintenance, National University of Shipbuilding, Elektronna Str. 61/39, 54031 Mykolaiv, Ukraine; dmytro.minchev@nuos.edu.ua
[5]   International Motor Bureau, Shkilna Str. 15, Nemishaeve, 07853 Kyiv Region, Ukraine; info@engine-expert.com
[6]   Department of Engineering Sciences, Danube Institute of National University "Odessa Maritime Academy", 9 Fanagoriyska Str., 68600 Izmail, Ukraine; zalozh@dinuoma.com.ua
*   Correspondence: kucera@fme.vutbr.cz; Tel.: +420-541-142-274

Citation: Varbanets, R.; Fomin, O.; Píštěk, V.; Klymenko, V.; Minchev, D.; Khrulev, A.; Zalozh, V.; Kučera, P. Acoustic Method for Estimation of Marine Low-Speed Engine Turbocharger Parameters. *J. Mar. Sci. Eng.* **2021**, *9*, 321. https://doi.org/10.3390/jmse9030321

Academic Editor: María Isabel Lamas Galdo

Received: 26 February 2021
Accepted: 10 March 2021
Published: 14 March 2021

Abstract: The article presents the acoustic method of marine low-speed engine turbocharger parameter estimation under operating conditions when a prompt assessment of instantaneous turbocharger speed and rotor vibration level is required. The method lies in the analysis of the acoustic signal that is generated by the compressor of the turbocharger with the diesel engine running under load. The spectral analysis reveals that the compressor blades generate acoustic oscillations that are always present in the overall acoustic spectrum of the turbocharger regardless of its technical condition. The harmonic components corresponding to the blades can be detected in the spectrum using the limit method. The calculated instantaneous turbocharger speed makes it possible to analyze the main harmonic amplitude in the spectrum. The method presented in this paper helps eliminate discrete Fourier transform (DFT) spectral leakage so that the amplitude of the main harmonic can be estimated. Further analysis of the amplitude of the main harmonic allows for efficient estimation of the turbocharger rotor vibration level when in operation. The method can be practically applied by means of a smartphone or a computer that has the dedicated software installed. The proposed method lays the foundations for a permanent monitoring system of turbocharger speed and vibration in industrial and marine diesel engines.

Keywords: turbocharger diagnostics; large diesel engine; acoustic spectral analysis; DFT leakage

1. Introduction

Modern turbochargers (TC) of large diesel engines enjoy a high boost pressure ratio in the compressor of up to 5 and above. They create high pressure of the charged air, thus providing high-specific power and high-efficiency operation of the large engine with low-level emission of carbon oxides and soot [1]. High efficiency of MAN MC (this type of MAN engine uses a mechanically driven camshaft for fuel injection) and MAN ME (electronically controlled engines) diesel engines with actual specific fuel consumptions of 160–170 g/kWh is ensured by the high pressure of the charged air, among other factors [2,3]. If the turbocharger loses performance, the power and efficiency of the diesel engine rapidly decline and the emission level of carbon oxides and soot increases [1–3].

The allowable hazardous emission level of marine diesel engines in operation is limited by the current requirements of the International Maritime Organization (IMO) [4].

Since the overwhelming majority of various maritime transport vessels use diesel power units, the matter of their efficient and safe operation is undoubtedly of current interest [5–7].

When a marine diesel engine is operating in light-load conditions, the incomplete combustion products clog up the exhaust manifolds. This results in a change in the flow capacity of the exhaust manifolds as well as the character of the gas internal flow in front of the turbocharger wheel blades. Pulsations might occur, which cause the rotor to vibrate [8–10]. An increased level of rotor vibration creates additional loads on turbocharger bearings and reduces their operational life. In the case of microdefects in the bearings, the vibration level of the rotor increases even further, which might lead to a severe failure [11].

Acoustic control of the turbocharger in operation makes it possible to detect the dangerous tendency for the rotor vibration level to increase and indicates the need to clean the flow channel [8–11]. In some cases, such control might prevent turbocharger failure, which typically leads to a considerable loss of power and efficiency in the entire engine [1,11–14].

Many authors have previously pointed out the necessity of conducting periodic operational checks of the technical condition of the turbocharger in operation [11–14]. In such an event, prompt and timely diagnostics during operation can be made by analysis of the external acoustic signals. In papers [11,12] it was indicated that within the spectrum of acoustic oscillations of the turbocharger, regardless of its technical condition, a compressor wheel blade frequency harmonic is always present. The amplitude of the blade harmonic of the compressor considerably exceeds (two, three times or even more) the level of surrounding harmonics in the turbocharger spectrum [8,12]. Meanwhile, the harmonic of the main rotation frequency of the turbocharger rotor might have insignificant amplitude and might not be distinguishable against the noises of the spectrum [8]. The turbine side blade frequency has generally significantly lower amplitude because of the much more stable and smooth confuser-type gas flow. For the large turbochargers with axial-type turbines, which have a number of blades three to five times greater than the compressor wheel, it is impossible to make a mistake between the compressor and turbine blade harmonics. Thus, the compressor blade harmonics are the primary source of the turbocharger spectrum analysis and can be identified in the spectrum using the method of limits.

This paper presents the diagnostic method based on the determination of the blade harmonics in the turbocharger spectrum, further calculation of the main rotor speed, as well as the subsequent analysis of the harmonic amplitude at its main frequency. The amplitude of the main frequency harmonic characterizes the general vibration level of the turbocharger rotor [5,6,15].

The article begins with an analysis of the acoustic signals of the TCA 66 and the VTR 564 turbochargers of low-speed diesel engines during their normal operation. The limits of normal and abnormal levels of the amplitude of the main harmonic are illustrated by these two examples. These cases, however, are specific, so it is necessary to analyze the acoustic spectrums of a larger number of turbochargers to develop general recommendations. Nevertheless, we believe that the information will be useful for marine engineers since these engines are widely used in merchant marine fleets.

To make the results of our analysis more reliable, the algorithm to eliminate the leakage effect is proposed, which helps to calculate more accurately the frequency and amplitude of the given signal.

A functional diagram of the continuous turbocharger monitoring system is proposed as the general conclusion of the paper.

The method in question can be applied practically; to do so in most cases, it would suffice to have a smartphone or a personal computer with the dedicated software. The method lies in the analysis of the acoustic signal that is generated by the turbocharger compressor while the diesel engine is operating under load.

2. TCA 66 Turbocharger of Low-Speed Diesel Engine Acoustic Analysis

The registration and analysis of acoustic signals of the TCA 66-20072 turbocharger [3], which is installed on the MAN 5S60MC main diesel engine [2] (Figure 1), was made at an engine speed of 85 rpm. The estimated engine brake power was $\approx$ 4500 kW or 50% of MCR (Maximum Continuous Rating), defined as the maximum output of an engine. According to the sea trial results, the corresponding turbocharger speed should be about 10,300 rpm (see Table 1). The compressor impeller wheel of the TCA66-20072 has 22 blades (11 full blades and 11 splitter blades) [3].

Figure 1. Engine 5S60MC and TCA 66-20072 turbocharger, 1—compressor wheel, 2—turbine rotor, (**a**,**b**)—acoustic measuring points.

Table 1. The main engine 5S60MC official test data.

Load	MCR	Engine	TC_{rmp}	TC_{in}	TC_{out}	P_{scav}	P_{max}	P_{comp}	SFOC
kW	%	rpm	rpm	°C	°C	bar	bar	bar	g/kWh
2208	25.0	66.1	6150	270	230	0.38	63.0	43.0	177.58
4407	49.9	83.0	9880	300	220	1.06	97.4	66.6	172.99
6621	75.0	95.4	12,050	320	200	1.78	129.8	95.0	168.16
7937	89.9	101.5	13,120	350	210	2.28	139.8	110.2	169.72
8824	99.9	105.3	13,850	372	220	2.62	139.8	124.0	171.13
8820	99.9	105.1	13,850	375	220	2.62	139.6	123.8	171.15
9673	109.5	108.2	14,540	410	240	2.96	140.0	136.0	172.50

Mechanical engineers provide engine performance analysis during its operation to estimate current engine and turbocharger conditions. The results of the performance analysis should be compared with manufacturer's official test data. Among the most important engine parameters for diagnostics are crank speed and brake power, brake specific fuel consumption, supercharged air pressure, compression pressure and maximum in-cylinder pressure, exhaust gas temperature at the turbocharger inlet and outlet, and turbocharger speed (Table 1 and Figure 2). According to the classifying societies requirements, the main engine should operate normally at 110% of maximum continuous rating (MCR) during 1 h.

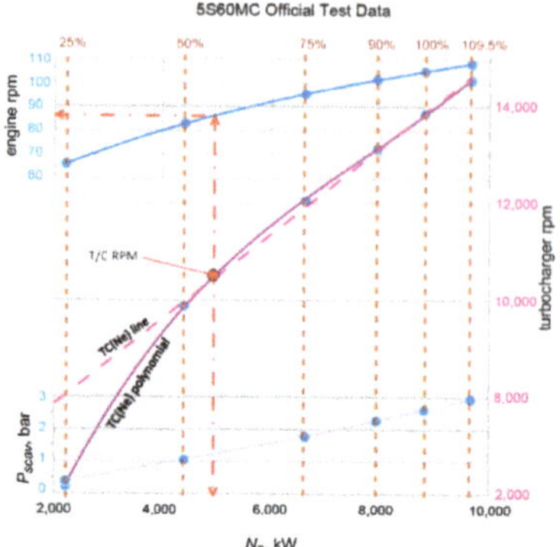

Figure 2. The main engine 5S60MC official test data ($engine_{RPM}$, TC_{rpm}, P_{scav}).

Deviations from these parameters for a given engine operating point may be caused by some failures. Thus, for instance, a compression pressure P_{comp} drop may be caused by cylinder liner wear, piston rings breaking or sticking, exhaust valve seat leakage/wrong timing, or by smaller supercharged air pressure P_{scav}. The reasons for a maximum combustion pressure P_{max} decrease are generally related to the high-pressure fuel-injection equipment. Smaller exhaust gas temperature differences at the turbine inlet and outlet ΔTC together with turbocharger speed TC_{rpm} reduction indicate flow duct fouling (Table 1).

$$\Delta TC = TC_{in} - TC_{out} \tag{1}$$

The suggested method allows the ongoing turbocharger speed to be measured at engine operation. The acoustic compressor signal registration could be carried out either at the compressor inlet filter (Figure 1a) or directly at the compressor volute surface (Figure 1b), as was learned from the experiment. The second option (point b) benefits from the inlet flow aerodynamic noise absence.

The aerodynamic noise could also be removed from the recorded signal if the microphone (positioned at point a) has a perpendicular orientation to the turbocharger intake filter surface, thereby providing smooth air flow around the microphone. The influence of the noise from other engine mechanisms is relatively small, as their sources generally have a big enough distance from the microphone position, as was learned from the set of experiments carried out on the number of low-speed marine engines.

At engine operation the compressor impeller blades generate oscillations in the general spectrum of vibrations regardless of the turbocharger's technical conditions, as was experimentally proven [5,8,11]. Spectral analysis shows that the acoustic signal from the compressor blades has a frequency equal to the turbocharger rotor speed multiplied by the number of blades (see Figure 3):

$$v_b = \frac{n_b TC_{rpm}}{60} \tag{2}$$

were v_b—blade frequency of the turbocharger, Hz; n_b—total number of compressor wheel blades; and TC_{rpm}—turbocharger speed, rpm.

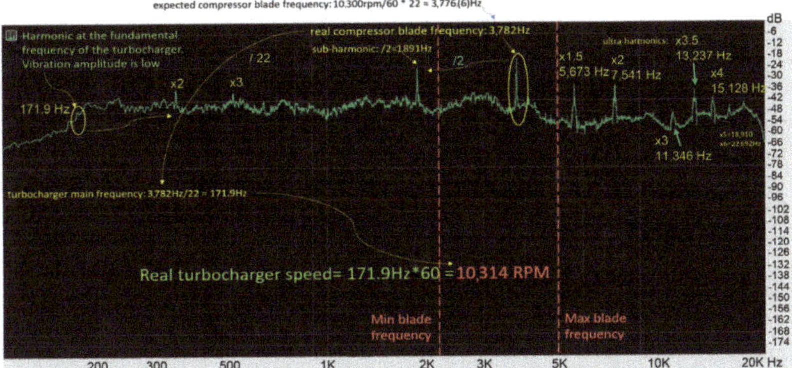

Figure 3. The acoustic signal spectrum of the TCA 66–20072 turbocharger compressor at 50% load mode of the 5S60MC engine (4500 kW, 85 rpm).

Assuming that current engine operating point is between 100% and 25% of MCR, the top and bottom limits for the turbocharger speed could be estimated (for the MAN 5S60MC engine):

- Max blade frequency $v_{bmax} = n_b \times \max(TC_{rpm})/60 = 22 \times 13{,}850/60 = 5078$ Hz,
- Min blade frequency $v_{bmin} = n_b \times \min(TC_{rpm})/60 = 22 \times 6150/60 = 2255$ Hz.

Thus, the blade harmonic in the general turbocharger spectrum lies between these limits (Figure 3):

$$v_{b\ min} < v_b < v_{b\ max} \tag{3}$$

The experimental measurements (Figure 4) were made with the electret microphone EM-4015-BC produced by Soberton Inc., Minneapolis, USA [16]. The microphone has high sensitivity, a wide pass band, a narrow directional pattern, little distortion, and a low noise level. It should be noted that due to the current microphone upper limit, the presented spectrum above 12 kHz could be incorrect, but it is not significant, as for low-speed diesel engines turbochargers it is not necessary to record signals over 10 kHz, so it does not affect the conclusion.

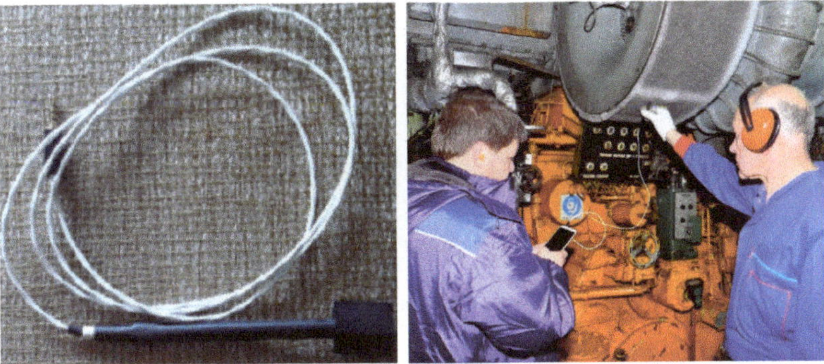

Figure 4. Recording the vibration of the turbocharger using the EM-4015-BC electret microphone.

The spectrum analysis (from Figure 3) revealed the value of the actual blade frequency: 3782 Hz. Thus, the actual turbocharger speed is equal to $3782/22 = 171.9$ Hz. Therefore, for a given engine operating point the TCA 66-20072 turbocharger actual speed is equal to $171.9 \times 60 = 10{,}314.5$ rpm.

It is notable that the main frequency (171.9 Hz) harmonic amplitude was relatively small and could be estimated at the range of spectrum noise. This could be assumed to be the indication of a small turbocharger rotor vibration level and highly probable as being the indication of normal rotor bearing conditions [8,12].

The acoustic spectrum of the TCA 66-20072 turbocharger also had sub-harmonics and ultra-harmonics:

- sub-harmonic $x0.5$ = 1891 Hz; and
- ultra-harmonics $x2$ = 7564 Hz, $x3$ = 11,346 Hz, $x4$ = 15,128 Hz, $x5$ = 18,910 Hz, $x6$ = 22,692 Hz.

Sub-harmonics and ultra-harmonics could be used as additional diagnostic signs for further experimental measurements of turbocharger operation.

For the signal, recorded at 44.1 kHz frequency, it was possible to conduct a spectrum analysis for harmonics with a frequency up to 22.05 kHz [17]. For most marine engines the frequency of the turbocharger blades is always at least 2 times smaller. The maximum frequency of the recorded signal is also limited by the microphone characteristics. As the possible step for spectrum analysis is down to 1 Hz, the absolute error of turbocharger frequency estimation is generally < 1 rpm. The measurement time should be related to the response time of the turbocharger, which is typically about 1 to 3 s. Therefore, the accuracy of suggested method for turbocharger frequency measurement exceeds the typical accuracy of standard measuring devices.

The suggested method could be used for accurate measurements of instantaneous turbocharger speed, turbocharger rotor vibration level estimation, and rough estimation of the engine brake power.

3. VTR 564 Turbocharger of Low-Speed Diesel Engine Acoustic Analysis

Another set of experimental research was carried out for the ABB VTR 564-31 turbocharger [18], installed on the MAN 6L80MC main marine engine of the capsize bulker [19].

The compressor impeller has 20 blades and its acoustic spectrum for the engine operating point close to MCR is shown in Figure 5. The assumed value of turbocharger speed for such conditions should be smaller than the turbocharger speed at the rated engine power. Therefore, the turbocharger speed for the MCR operating point could serve as the top limit.

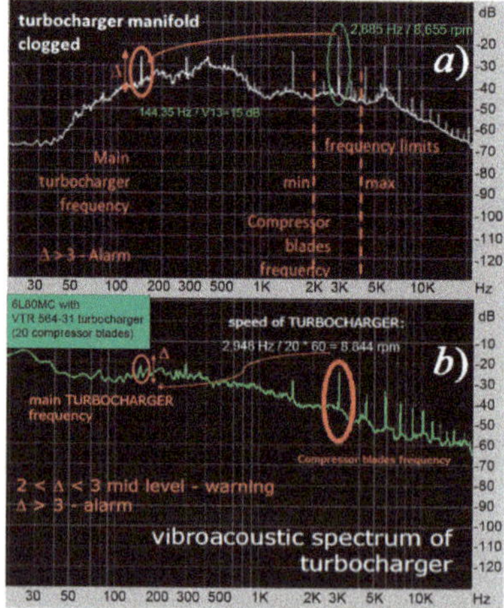

Figure 5. Acoustic spectrum of the VTR 564 turbocharger before (**a**) and after (**b**) manifold cleaning.

Max blade frequency v_{bmax} = nb × max(TC_{rpm})/60 = 20 × 12,000/60 = 4000 Hz.

As the engine operating point could be estimated as close to the MCR or at least between 50% and 100% of the MCR, the bottom limit of the turbocharger speed could be assumed to be v_{bmin} = 2000 Hz.

For this turbocharger two cases were considered: before and after cleaning the engine exhaust system. The corresponding acoustic spectra are presented in Figure 5.

The measurements made on the engine before the exhaust system cleaning gave a main harmonic frequency equal to 144.35 Hz, which was calculated from the blade frequency of the spectrum. It is clear that a main harmonic amplitude greater than 15 dB exceeded the general spectrum level (Figure 5a). This increased level of the main harmonic amplitude was caused by the vibrations of the turbocharger rotor at the frequency of its rotation. The reason was dust, oil, and carbon deposits on the turbine nozzle and exhaust manifold walls, which distorted the gas flow path and caused vibrations in the turbocharger rotor.

Cleaning and revision of the exhaust manifold and turbine side of the turbocharger allowed the main harmonic amplitude to be lowered down, as shown in Figure 5b. It is obvious that the level of the main harmonic amplitude was equal to the signal noise level in this case.

By scaling the spectrum diagram (Figure 5b), the value of blade frequency was estimated as 2948 Hz—it was the closest harmonic to the top limit of 3 kHz. The left harmonic in the diagram (respective to the blade harmonics) is the sub-harmonic, which had a frequency two times smaller, 1474 Hz.

From the turbocharger blade frequency, the turbocharger rotor speed was calculated as TC_{rpm} = 60 × 2948/20 = 8844 rpm.

The turbocharger standard tachometer indicated the turbocharger speed as 8800 rpm, so the error of the turbocharger speed estimation was about 0.5%. It is important to underline that as the error of the turbocharger speed estimation by the acoustic method is generally less than 1 rpm, it provides much more accurate turbocharger speed and further engine operating point estimation.

The frequency of the main harmonic for the turbocharger rotor speed was $v_{turbocharger}$ = v_b/n_b = 2948/20 = 147.4 Hz.

Obviously, the relatively high level of the main harmonic amplitude Δ could be an indication of an increased level of rotor vibrations [8]. The level of the main harmonic amplitude Δ in Figure 5 could be assumed to be lightly increased but still permissible. The relative main harmonic amplitude parameter δ is proposed to assess the level of Δ:

$$\delta = \frac{\Delta}{\sqrt{D}} \tag{4}$$

where D is the variance of the signal level, calculated in a range from $v_{turbocharger}$ − 50 Hz to $v_{turbocharger}$ + 50 Hz and excluding the level of the main harmonic.

In all cases of measurements on marine engines, when high levels of fundamental harmonics were found, it was necessary to clean the flow path of the turbochargers. After cleaning, the level of the fundamental harmonic decreased down to δ < 2.

The set of experiments carried out on the number of MAN MC-series diesel engines shows that if the main harmonic amplitude becomes two to three times greater than the average spectrum level, it indicates a dangerous level of turbocharger rotor vibrations [12].

More precise quantitative assessment of the permissible range of turbocharger rotor vibrations level requires further experimental investigations. Among other factors, it depends on the design peculiarities of the particular turbocharger: type and positioning of rotor bearings (sleeve or roller type, interior or external position), type of the exhaust gas supply to the turbine wheel (pulsating or constant-pressure flow, single or multiple inlet), and the type of turbine (axial or radial), etc. It is important to emphasize the ability to make such an assessment rapidly during normal engine operation and without any additional devices installed on the engine.

4. Eliminating the Leakage Effect of the Discrete Spectrum (LEE)

When analyzing the discrete spectrum of acoustic signals to estimate their frequency and amplitude properties, the problem of leakage effect elimination must be solved. This effect arises due to the finiteness of the time analysis carried out and its discrete representation. The leakage effect from spectral peaks to adjacent spectral lines is considered to be one of the main errors of the DFT [17].

As an example, Figure 6 shows the amplitude spectra of an identical sine signal with an integer (a) and a non-integer (b) sample number during one signal period.

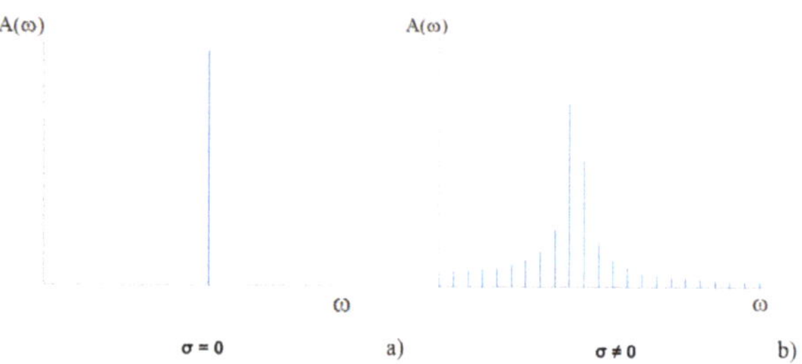

Figure 6. The DFT leakage effect. Integer (**a**) and a non-integer (**b**) sample number during one period.

Let the frequency of a signal be represented by

$$\gamma = \frac{M}{T} \tag{5}$$

where T is the period of the signal and $M = n + \sigma$, wherein n is an integer and $0 < \sigma < 1$.

The maximum distortion of the amplitude, frequency, and phase of the central harmonic and the power leakage into adjacent spectral components occurs at $\sigma = 0.5$ [17].

Thus, when analyzing the parameters of the original signal of the spectrum, i.e., the central harmonics, the calculated amplitude, frequency, and phase will be set incorrectly if the number of samples per period is not an integer. In technical practice, an analog-to-digital converter (ADC) with a selected and fixed sampling frequency is used for discrete signal recording. It is therefore logical that the sample number during one period could never be integer, the σ parameter can lie in the interval from 0 to 1 depending on the measured signal and eigenfrequency, and the accuracy of estimating the signal parameters along the central harmonic will change.

The most common solution for leakage effect reduction is based on window transform methods. The principle of this method is to reduce the number of discontinuities at the edges. To reduce leakage it is necessary to reduce the signal amplitude of the signal near the edges. This procedure is carried out using the multiplication by the window function with the special form $s_j^w = s_j W(j)$, where $W(j)$ stands for window functions (Hamming, Hanning, Kaiser, etc.).

Using the time window, the original signal spectrum is changed and its amplitude is reduced by RMS Coeff times. This reduces the dependence of the basic harmonic amplitude on the value σ. This procedure allows the fundamental harmonic to be used to approximate the parameters of the signal with constant error, which is already acceptable for the intended purpose.

More specifically, the leakage effect can be eliminated by using a numerical method that processes complex DFT results. In [17], a suggestion was made that the frequency m, the phase φ, and the amplitude A of the original signal from the values of two maximum harmonics in the spectrum should be specified. To do this, it is necessary to numerically solve a system of equations in a complex variable numerically:

$$\left\{ \begin{array}{c} \left| \dfrac{E(m,\phi)_k}{E(m,\phi)_{k+1}} \right| = \left| \dfrac{X_k}{X_{k+1}} \right| \\ arg(E(m,\phi)_k) = arg(X_k) \end{array} \right\} \tag{6}$$

where the *k*th harmonic parameters are specified as

$$X_k = Re_k + jIm_k \tag{7}$$

$$X_k = NA_k e^{j\varphi_k} \tag{8}$$

$$A_k = \frac{1}{N}\sqrt{Re_k^2 + Im_k^2} \tag{9}$$

$$\phi_k = \arctan\left(\frac{Im_k}{Re_k}\right) = arg(X_k) \tag{10}$$

The harmonic coefficients can be represented in the form:

$$X_k = \left(\frac{A_k}{2}\right)E(m,\phi)_k \tag{11}$$

where $E(m,\phi)_k$ is a complex function independent of the amplitude, but dependent on the frequency and phase:

$$E(m,\ \phi)_k = e^{j\phi}\frac{e^{2\pi j(m-k)}-1}{e^{\frac{2\pi j(m-k)}{N}}-1} + e^{-j\phi}\frac{e^{-2\pi j(m-k)}-1}{e^{\frac{-2\pi j(m-k)}{N}}-1} \tag{12}$$

The method of solving Equation (6) was developed by R. Varbanets. He revealed the influence of noise level in the source signal on the calculated amplitude and frequency of the main harmonic (see Figure 7, green line).

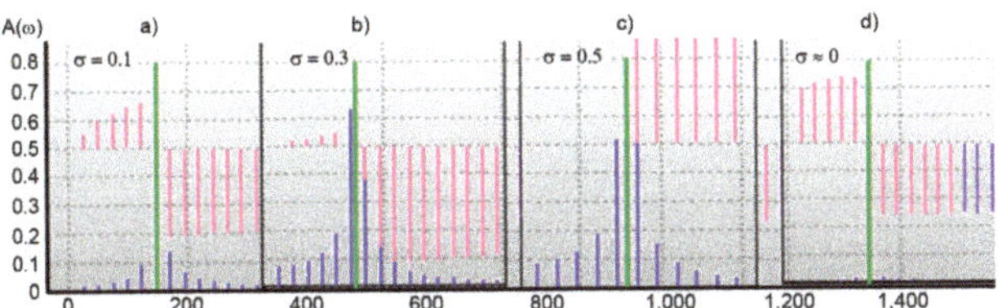

Figure 7. DFT leakage effect elimination by solving Equation (6), (**a**) σ = 0.1, (**b**) σ = 0.3, (**c**) σ = 0.5, and (**d**) σ = 0.

Equation (6) needs to be solved in case the lateral harmonic components are different from zero, i.e., greater than a specified small value ε:

$$X_{k-1} > \varepsilon \text{ and } X_{k+1} > \varepsilon \tag{13}$$

If $X_{k-1} = 0$, $X_{k+1} = 0$, then the leakage effect is suppressed and the frequency, amplitude, and phase of the central harmonic correspond to the initial measured signal parameters (Figure 6a).

'When solving Equation (6) for the situation of strong leakage effects (σ ~ 0.5), after five iterations an error of less than 0.5% in frequency and phase was achieved. For a sinusoidal signal, the amplitude and frequency were restored to the values contained in the original record to 5 decimal places. Prior to this recovery procedure, the central harmonic error was 35% [12,17].

In addition, the error of estimating the frequency of the original record can be not negligible in relation to the frequency of the central harmonic. As the ADC sampling rate increases, this error decreases.

The solution to Equation (6) is not associated with additional memory, as is the case for the fast Fourier transform (FFT). Despite the iterative numerical solution for Equation (6), such a procedure only very slightly increases the overall computation time and makes it possible to obtain not only the spectrum of the signal, but also the restored fundamental frequency value, amplitude, and phase of the measured signal when it is close to sinusoidal.

This method was investigated in the case of noise in the original signal (with a white noise of 5% and 10% of the amplitude of the sinusoid). Figure 7 shows the solution of Equation (6) for a sinusoid with an amplitude of 0.8 and for the cases (a) $\sigma = 0.1$, (b) $\sigma = 0.3$, (c) $\sigma = 0.5$, and (d) $\sigma = 0$.

The central green line in each Figure 7a–d is the main harmonic of a sinusoid with an amplitude of 0.8 with the restored amplitude, frequency, and phase, the result of solving Equation (6).

In all solved cases, no more than 5 complete iterations were needed to ensure the given accuracy. As a result of the solution of Equation (6), the phase and frequency of the signal with the addition of white noise to 10% were restored to the initial value, with an error of not more than 0.5%.

5. Application of the Acoustic Method for a CNG Engine

The described acoustic method was also successfully applied to a completely new engine developed for small vessels or locomotives, where compressed natural gas (CNG) is used instead of traditional fuels (see Figure 8). The measurement was performed in the same configuration as described above for the engine shown in Figure 4.

Figure 8. CNG engine and its application in an experimental rail vehicle.

This CNG engine contributes to reduced environmental impact and has low operating costs. The use of these high-volume engines is not only in shipping and rail transport, but also in construction machinery and cogeneration units. The possibilities of using these engines are really wide thanks to ecological aspects, low price, and rich reserves of natural gas all over the world.

6. Acoustic-Based Turbocharger Continuous Monitoring System

The experiment was carried out with a smartphone and external microphone, with further analysis of the recorded acoustic signals carried out on a personal computer. The blade frequency and main harmonic amplitude on the turbocharger rotor frequency were estimated after signal spectrum calculations and DFT leakage effect elimination. As the signal measurement and treatment requires very little time, the stationary system for continuous turbocharger acoustic monitoring could be proposed.

The suggested system should include a special industrial-class microphone, installed either close to the compressor blades or outside the compressor volute, in a place with minimum interaction from aerodynamic noises or noises from engine mechanisms. The microphone should be a directional type and orientated to the signal source. We used a special protector made from porous rubber to cover the microphone from extraneous noise (see Figure 4).

The signal from the microphone was processed by the DSP controller, which implements FFT, spectrum build-up, and DFT leakage effect elimination (Figure 9). The instantaneous turbocharger rotational speed at some quasi-steady time interval and the relative rotor vibration level δ were the output parameters.

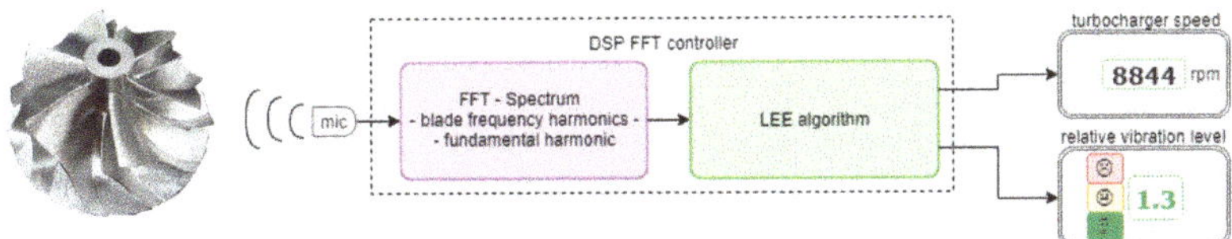

Figure 9. Block diagram of the acoustic-based turbocharger continuous monitoring system.

The quasi-steady time interval depends on the turbocharger inertia. It is assumed that the process at the time interval, which is shorter than the turbocharger time constant, could be considered a steady process, and the FFT procedure could be applied. For low-speed turbocharger engines the time interval is generally less than 1 s.

7. Conclusions and Further Research

In accordance with the main directions of increasing the efficiency of diesel engines it is relevant to reduce operating costs by using more technologically advanced and practical equipment [20,21]. The suggested system for continuous acoustic turbocharger monitoring could increase the efficiency of low-speed marine engine maintenance. At the same time the method based on a smartphone application could be a useful tool for marine engineers, as it does not require special equipment (in most cases in-built-in microphone could be used).

To define the limits of normal and abnormal rotor vibration levels for various types of turbochargers, further research is necessary. It can be noted that the acoustic spectrum analysis of a turbocharger can be quickly made under operating conditions.

The proposed method allows the turbocharger rotor speed and the vibration level to be determined by means of acoustic spectrum harmonic amplitude stabilization using the proposed algorithm. The method can be implemented in the continuous monitoring system of the turbocharger (Figure 9). Tests on two-stroke marine engines have shown that the acoustic-based measuring system can be carried out, which allows:

- continuous monitoring of turbocharger rotation speed and rotor vibration level;
- reliability and ease of installation, as the sensor is in a low temperature zone; and
- high accuracy rotation speed control, which makes it possible to monitor the total engine load.

The described acoustic method for estimating the parameters of turbochargers has already proved to be a useful tool for the development and operation of engines in its cur-rent form. The major advantage of the acoustic method is its simplicity. There is no need in expensive high-quality vibration sensor, its installation and calibration. Generally, even regular smartphone with in-build microphone can provide good results. For correct application, new users of this method may be advised to first determine the frequency spectrum of the relevant turbocharger in its nominal operating conditions.

During the development and application of this measuring method, the influence of the acoustic background was also investigated, for example due to the running of another engine running nearby. This problem can be solved applying the directional microphone and its positioning close to the source of the blade harmonics sound, as has been success-fully verified.

This method can also be used for automotive engines but requires some modifications. First, the microphone should be able to record sound up to 40 kHz and second, special software should be able to record sound at 80.2 kHz. When using a data rate of 44.1 kHz, the maximum detectable turbocharger speed is approximately 70,000–100,000 rpm.

Further research and development of the continuous monitoring system is focused on the automatic detection of excessive turbocharger rotor vibrations and the appropriate intervention without an operator. The development of appropriate algorithms and software is underway, which will be the content of another publication.

The extension of the application of the method to the issue of detection of transient modes of the turbocharger is also considered. The time constant of the large marine engine turbochargers could be estimated as 0.5–2 s. So, if the turbocharger speed is relatively constant within this period of time, the method could be applied even for engine transient op-eration. For faster transition processes, special processing of the recorded signal is required, which is the subject of ongoing research.

Author Contributions: Conceptualization, R.V. and V.K.; methodology, R.V., O.F., and V.K.; software, R.V., V.K., and D.M.; validation, A.K. and V.Z.; writing—review and editing, V.P., P.K., A.K., and V.K. All authors have read and agreed to the published version of the manuscript.

Funding: The authors gratefully acknowledge funding from the specific research on BUT FSI-S-20-6267.

Institutional Review Board Statement: Not applicable.

Informed Consent Statement: Not applicable.

Data Availability Statement: Not applicable.

Acknowledgments: The authors thank Odessa National Maritime University and National University of Shipbuilding, Mykolaiv, for support. The authors thank the IMES GmbH company for providing the data sets employed in this work. The authors thank Brno University of Technology for their support.

Conflicts of Interest: The authors declare no conflict of interest.

Abbreviations

DFT	Discrete Fourier Transform
FFT	fast Fourier Transform
LEE	leakage effect elimination
ADC	analog-to-digital converter
TC	turbocharger
IMO	International Maritime Organization
SFOC	specific fuel oil consumption
MCR	maximum continuous rating
CNG	compressed natural gas
P_{max}	maximum combustion pressure
P_{scav}	scavenging air pressure
P_{comp}	pressure at the end of compression
TC_{rpm}	turbocharger rotor speed
TC_{in}	exhaust gas temperature at the turbocharger inlet
TC_{out}	exhaust gas temperature at the turbocharger outlet
n_b	compressor wheel blade number

References

1. Heywood, J.B. *Internal Combustion Engine Fundamentals*; McGraw-Hill Education: London, UK, 2018.
2. MAN B&W S60MC-C8.2-TII Project Guide. Available online: https://marine.man-es.com/applications/projectguides/2stroke/content/printed/S60MC-C8_2.pdf (accessed on 9 January 2021).
3. TCA Turbocharger Project Guide. Available online: https://turbocharger.mandieselturbo.com/docs/default-source/shopwaredocuments/tca.%20pdf?sfvrsn=2 (accessed on 9 January 2021).

4. Čampara, L.; Hasanspahić, N.; Vujicic, S. Overview of MARPOL ANNEX VI regulations for prevention of air pollution from marine diesel engines. In Proceedings of the SHS Web of Conferences, Samara, Russia, 26–27 November 2018; Volume 58, pp. 1–10. [CrossRef]

5. Varbanets, R.; Karianskiy, A. Marine diesel engine performance analyze. *J. Pol. CIMAC* **2012**, *7*, 269–275.

6. Fomin, O.; Lovska, A.; Píštěk, V.; Kučera, P. *Dynamic Load Computational Modelling of Containers Placed on a Flat Wagon at Railroad Ferry Transportation*; Vibroengineering Procedia; Greater Noida: Delhi, India, 2019; pp. 118–123. [CrossRef]

7. Fomin, O.; Lovska, A.; Píštěk, V.; Kučera, P. Research of stability of containers in the combined trains during transportation by railroad ferry. *MM Sci. J.* **2020**, 3728–3733. [CrossRef]

8. Solomatin, S. *Foundations of Technical Diagnostics*; ONMU: Odessa, Ukraine, 2007; p. 80. (In Russian)

9. Píštěk, V.; Kučera, P.; Fomin, O.; Lovska, A. Effective mistuning identification method of integrated bladed discs of marine engine turbochargers. *J. Mar. Sci. Eng.* **2020**, *8*, 379. [CrossRef]

10. Píštěk, V.; Kučera, P.; Fomin, O.; Lovska, A.; Prokop, A. Acoustic identification of turbocharger impeller mistuning—A new tool for low emission engine development. *Appl. Sci.* **2020**, *10*, 6394. [CrossRef]

11. Zigelman, E.; Skvortzov, D.; Loshinin, I. Study of possibility for vibrodiagnostics of medium diesel generators. *Izv. Vuzov* **2013**, *6*, 42–48. (In Russian)

12. Varbanets, R.; Kucherenko, Y. Turbocharged marine diesel engine frequency parameters monitoring. *Bull. Astrakhan State Tech. Univ. Ser. Mar. Equip. Technol.* **2013**, *1*, 103–110. (In Russian)

13. Kostyukov, V.; Naumenko, A. Condition monitoring of reciprocating machines. In Proceedings of the 22nd International Congress on Condition Monitoring and Diagnostic Engineering Management: COMADEM 2009, San Sebastian, Spain, 9–11 June 2019; pp. 113–120.

14. Naumenko, A. Real-time condition monitoring of reciprocating machines. In Proceedings of the 6th International Conference on Condition Monitoring and Machinery Failure Prevention Technologies 2009, Dublin, Ireland, 23–25 June 2009; pp. 1201–1212.

15. ISO. Standard ISO 10816. In *Mechanical Vibration—Evaluation of Machine Vibration by Measurements on Non-Rotating Parts*; ISO: Geneva, Switzerland, 2014.

16. EM-4015-BC, Analog Microphone Electret Condenser 1V~10V Omnidirectional (−44dB ±3dB @ 94dB SPL) Solder Pads. Available online: https://www.soberton.com/em-4015-bc/ (accessed on 5 November 2020).

17. Otnes, R.K.; Enochson, L. *Applied Time Series Analysis*; Wiley: New York, NY, USA, 1978; p. 428.

18. VTR564E32 ABB Turbo Systems. Available online: https://library.e.abb.com/public/18a4237f8f5b406e9a9a92aa74aeb501/ZTL2 104.pdf (accessed on 9 January 2021).

19. Hanjin Dampier. Navigation Act 1912 Navigation (Marine Casualty) Regulations Investigation into the Grounding of the Korean Flag Bulk Carrier Hanjin Dampier at Dampier in Western Australia on 25 August 2002. Available online: https://www.atsb.gov.au/media/24941/mair184_001.pdf (accessed on 5 November 2020).

20. Puškár, M.; Kopas, M.; Sabadka, D.; Kliment, M.; Šoltésová, M. Reduction of the gaseous emissions in the marine diesel engine using biodiesel mixtures. *J. Mar. Sci. Eng.* **2020**, *8*, 330. [CrossRef]

21. Fomin, O.V. Increase of the freight wagons ideality degree and prognostication of their evolution stages. *Sci. Bull. Natl. Min. Univ.* **2015**, 68–76. Available online: http://nv.nmu.org.ua/index.php/en/monographs-and-innovations/monographs/1078-engcat/archive/2015/contents-no-3-2015/geotechnical-and-mining-mechanical-engineering-machine-building/3040-increase-of-the-freight-wagons-ideality-degree-and-prognostication-of-their-evolution-stages (accessed on 13 February 2021).

MDPI
St. Alban-Anlage 66
4052 Basel
Switzerland
Tel. +41 61 683 77 34
Fax +41 61 302 89 18
www.mdpi.com

Journal of Marine Science and Engineering Editorial Office
E-mail: jmse@mdpi.com
www.mdpi.com/journal/jmse